D0539271

THE SUN ISLANDS

By the same author:

A Natural History of the Cuckmere Valley, The Book Guild, 1997

The Mountain of Mist – a novel based on a true story, The Book Guild, 1998

THE SUN ISLANDS

A Natural History of the Isles of Scilly

Patrick Coulcher

The Book Guild Ltd
Sussex, England

The Book Guild Ltd,
25 High Street,
Lewes, Sussex

First published 1999

© Patrick Coulcher, 1999

Set in Times
Typesetting by Acorn Bookwork, Salisbury, Wiltshire

Origination, printing and binding in Singapore under the
supervision of MRM Graphics Ltd, Winslow, Bucks

A catalogue record for this book is available from the British Library

ISBN 1 85776 348 3

CONTENTS

LIST OF MAPS

LIST OF ILLUSTRATIONS

To my wife Margaret and to many friends who have shared with me the allure and splendour of The Islands.

ACKNOWLEDGEMENTS

I could not have written this book without the help of many, many people. Particularly I wish to thank those people of The Isles of Scilly, Robert Dorrien-Smith, Andrew Gibson (The Environmental Trust), Frank Gibson, Vivian Jackson, David Knight, Steve Ottery (Curator, Isles of Scilly Museum), Jeremy Pontin (Land Steward, Duchy of Cornwall), William Wagstaff and Steve Walder. My thanks to Betty Hamilton for allowing me to include some of her late husband's paintings; I knew John Hamilton well and frequently visited his studio when it was located on Tresco – he captures the mood and atmosphere of the Islands so perfectly in his work.

That great artist, the late Frank Wootton, OBE, accompanied me on a tour of The Islands just before he died and I am proud to include some of his work, particularly drawings of birds and seals. I am also indebted to another local artist, Sue Lewington whose sketches bring life to each chapter. My thanks also to Michael Hollings for his wonderful photographs, Roger Taylor for his help and Pat Donovan for her two beautiful illustrations of wild flowers. Thank you, Westmin Photographic of Eastbourne, who processed many of my photographs included in the book.

I am grateful for the assistance of friends like Sally and David Candlin, Joan and Geoffrey Stewart, Hilda Gallimore, and my sister Susan Millikin who have shared the delights of Scilly and its wildlife. I thank too my daughter Jane for her poem at the end of the book.

To Joy and Donald Preen my gratitude for your helpful comments on the manuscript and to Marion Wood my appreciation for your hard work in typing my scribbled words.

Finally, I wish to thank His Royal Highness The Prince of Wales for the great honour in writing the Foreword of this book.

ST. JAMES'S PALACE

The Isles of Scilly have always been a very special place for me. They have been the "jewel in the crown" of the Duchy of Cornwall's historic Landed Estates since 1337. Their unique environment needs to be cared for and cherished so that it can be passed on unspoilt to future generations, and it is for this reason that the Islands have been recognised as an Area of Outstanding Natural Beauty with twenty five Sites of Special Scientific Interest, set within a Marine Park.

I welcome Patrick Coulcher's new book as a contribution to wider understanding of the special importance of the Isles of Scilly. I am particularly grateful that the Isles of Scilly Environmental Trust, which was created in 1985 with the assistance of the Islands Council, will benefit from the contribution which some of the royalties of this book will make to their role in dealing with the pressures of modern man on a historic, precious and vulnerable landscape.

INTRODUCTION

I have always been attracted to islands and as a small boy I avidly collected books on those around the coast of Britain to learn more about their past and natural history. My first sight of The Isles of Scilly was in 1958 when, as a young pilot in the Royal Air Force, I flew Hunter aircraft on low-level training sorties from a base in north Devon. I discovered those beautiful jewels of land, their shades of green and brown blending in so splendidly with the pure white sandy beaches and ultramarine blue of the sea. They lie 28 miles west-south-west of Land's End.

It was many years later in the 1980s that I was able to visit and explore the islands at first hand. The more I experienced their charm and beauty the more I wanted to see and learn about their history and about their natural life, their birds, butterflies and wild flowers. Every year I visit them, sometimes in the spring and autumn, but usually in the early summer months of May and June when the birds have their nests and can easily be watched, the flowers are at their best, and the butterflies are about in large numbers taking their fill of the abundant nectar. For good reason these islands have, in the distant past, been known as the Sulli Islands or Sun Islands.

I decided to write this book in the hope that I can convey my feelings and appreciation of their charm to other people who may learn just a little more about them. The book starts with a brief history of man in Scilly which sets the scene for the more detailed natural history to follow. I have divided the area of the islands into ten individual groups or single entities and each has its own dedicated chapter which includes the natural species to be found there. Some species are peculiar to only one particular locality and in general this will be mentioned, but of course most will be found in many places right across the area.

Clearly this book is not intended to contain a full and precise description of every species on the islands; to do so would require an encyclopedia. Nor is it intended for the professional naturalist, although he or she might find something of interest in it. What this book does try to do is to describe some species which have beauty or interest or are peculiar to the area. As such, I hope it will appeal to those seeking knowledge about the natural world, and to those who just want to relax, perhaps at home, and read about the allure, beauty and tranquillity of this lovely part of the British Isles.

The reader is invited to explore the areas described in the chapters at leisure, using the many facilities available from scheduled boat trips to private arrangements made with local fishermen. Although I have included an overall map of the islands and each chapter has a general guide map, many more features will be found using an Ordnance Survey map such as the Explorer 20, 1:25,000 scale, entitled 'Isles of Scilly'. I have highlighted the books in the Bibliography which will be found useful as a guide to identifying the birds, butterflies and plants mentioned in the text. Do buy a good hand lens (× 10 magnification) to examine the detailed structure of plants, and you will not be disappointed by the sheer beauty revealed in the form and make-up of our wild flowers.

The Isles of Scilly situated within the warm Gulf Stream have an equable climate, the temperature seldom getting lower than about −2°C or going above 27°C. The rainfall of 32.7 inches per year is the same as the average for the whole of England, the driest months being from May to June. The relative humidity for the islands is high throughout the year and varies between 81% and 86%. These climatic factors all favour a wide variety of plant life and as we shall see, because of their position and association with worldwide merchant trade, many of the islands' wild flowers originated from far-flung places on the globe.

In addition to flowers, the islands, because of their geographical position, regularly attract a wide range of bird species from all regions of the northern hemisphere. As well as normal west coast migrant birds and those from eastern and southern Europe, the islands also receive a remarkable number of vagrant North American birds brought in by the prevailing westerly winds. So in your wanderings you could possibly see rare species of bird, especially during the spring and autumn migration periods. The islands too, because of their mild climate and diverse agricultural system, support breeding varieties of songbirds that are in excess of those found on the mainland. So you will see many thrushes, blackbirds, dunnocks, tits and finches on your walks, and some of them are remarkably tame.

By kind permission of The Isles of Scilly Museum I have added a checklist of birds (revised 1993) and also a list of flowering plants mentioned in the text, together with their Latin names. These checklists can be found at the back of the book.

The reader will wish to know right from the start that there are no snakes, foxes, weasels, stoats or lizards (except the slow-worm) on the Isles of Scilly and there are very few wasps. But rabbits, introduced as a source of fresh meat in the Middle Ages, do flourish on most of the islands.

I am very pleased that half the royalties received will be given to The Isles of Scilly Environment Trust, which was set up in 1985 by the Duchy of Cornwall to protect the islands' precious heritage for future generations to enjoy. The Trust's book, *A Precious Heritage*, details those places which are susceptible to disturbance, and have restricted access to members of the public (included at the end of this book). Apart from a marvellous description of the islands it also sets out a code of conduct for visitors.

Wherever you go and whatever you see, remember to protect and respect the countryside and environment and respect the peace, quietness and beauty that others have also come to enjoy.

WARNING: Medicinal properties of some plants and fungi are given in this book. In no circumstances should readers experiment or try these out themselves without sound professional advice.

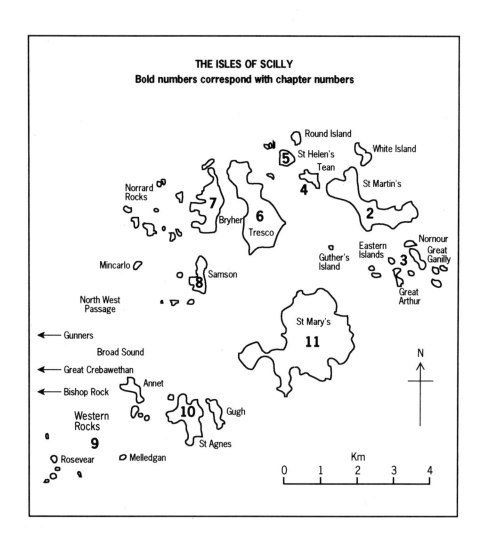

THE ISLES OF SCILLY
Bold numbers correspond with chapter numbers

1

A HISTORY OF THE ISLES OF SCILLY

Cromwell's Castle, Tresco

Prehistory and Legend

> Then rose the King and moved his host by night,
> And ever push'd Sir Mordred, league by league,
> Back to the sunset bound of Lyonesse –
> A land of old upheaven from the abyss
> By fire, to sink into the abyss again;
> Where fragments of forgotten peoples dwelt,
> And the long mountains ended in a coast
> Of ever-shifting sand, and far away
> The phantom circle of a moaning sea.
>
> Tennyson
> *The Passing of Arthur*

The Hesperides, Elysian Fields, Atlantis or simply the Fortunate
Islands; these are the names mentioned in Greek and Latin mythol-
ogy of a group of islands situated beyond the Pillars of Hercules
(Straits of Gibraltar). Here, it was said, the Greeks buried their
dead heroes where they would find immortality in a land filled with
meadows of wild flowers and where sunshine gave never-ending
warmth. This is not quite as far-fetched as it may seem, as it is
known that ancient civilisations did indeed bury their leaders on

1

islands where they would not return across the sea and interfere with their successors. One thing is fact, and that is there are a large number of burial mounds and monuments on the islands, possibly many more than could be justified by the number of people who actually lived there.

The legend of King Arthur also permeates the history of the islands. For centuries the lost land of Lyonesse has been described and explained by poets and writers as part of a large tract of land beyond Land's End. King Arthur, that Christian King of Britain who fought the pagan Anglo-Saxon invader, was said to rule over Devon, Cornwall and Lyonesse from his castle at Tintagel in Cornwall. Many Cornishmen still believe in the existence of a large town called 'The City of Lions' at the Seven Stones, some six miles north-east of St Martin's, now marked by a lightship. Indeed, it is said that many artefacts such as pieces of buildings, glass windows and utensils have been dredged up at different times from the spot still spoken of as 'The Town'. King Arthur of course was killed in battle in Cornwall and his followers, chased by their foe Prince Mordred, fled to Lyonesse (Scilly) and the land by which they passed then sank into the sea, engulfing their pursuers.

In your wanderings through and about the islands you will find many references to Lyonesse; then remember the legends of these ancient island peoples.

Early Man in the Scilly (10,000 BC–AD 43)

Over 10,000 years ago southern Britain was roamed by bands of Palaeolithic people. Little is known of their way of life, but their cultures covered a long period of time, representing man's evolution from ape to *Homo sapiens*. They sheltered in caves and were hunters and food gatherers with a nomadic or semi-nomadic existence, moving on when local food supplies ran out or to follow migrating animals. Few signs of their existence have been found on Scilly because they made easily biodegradable shelters of animal hides, using materials such as bone, wood and skin. Only their stone tools and implements such as arrowheads have not perished with time, and evidence of these have been found on the islands. For instance a flint-working site of this period has been identified at Old Quay on St Martin's.

During the Palaeolithic Age The Isles of Scilly would have looked very different than they do today. Then the seas were much lower with much of the planet's water locked up in ice sheets during the glacial period and the islands would have consisted of

2

one single land mass stretching from the Eastern Islands to the Western Rocks, and from the south of St Mary's to the north of Bryher. There is evidence that this land mass would have had meadows and sheltered slopes covered with mixed woods of oak, hazel, elm, ash and birch. There would have been many large animals such as red deer, and the surrounding seas would have been teeming with fish, seals, shellfish and edible seaweeds. Man would easily have been able to live and perpetuate in this land of plenty.

From about 8000 BC to 4000 BC, Mesolithic people, who were more settled, would have occupied Scilly. These people would have burnt out the virgin forest to flush out game and to create larger pastures for crop growing.

As the glaciers melted so the seas rose, and by 3000 BC the land mass of Scilly would have become a number of separate islands. The Neolithic Age had begun and from about 4000 BC to 2500 BC man had begun to establish more permanent settlements with complex social and political groupings. In Scilly, only a few stone axes and flint arrowheads, and some pieces of Neolithic pottery, have been discovered to mark this period. No ceremonial large stone monuments of this age have been discovered, but the submergence will have destroyed any of the low-lying sites, and since most surviving entrance graves are unexcavated, more evidence of Neolithic settlements is awaited.

It was the Bronze Age, from 2500 BC to 700 BC, that brought with it people from west Cornwall who permanently settled in Scilly. Many traces of their occupation exist today, including houses, flint instruments, stone mortars and, most numerous of all, their ceremonial monuments. Most impressive are their entrance graves, which exist on most of the islands and consist of a circular stone and earth cairn containing a rectangular chamber covered by several large capstones. Some of these graves are entered by an uncovered entrance passage, much as the one at Bant's Carn on St Mary's. Standing stones (menhirs) from this period also exist and some can be seen at Chapel Down on St Martin's. Bronze Age people had a mixed subsistence economy, growing crops of barley and wheat and keeping livestock, as well as hunting for wild animals and birds. Fishing was an important industry, of course, and fish were preserved by wind-drying and salting from evaporated brine. Shellfish of all kinds were used as bait and for food, and their waste tips or middens full of empty shells can still be found on such places as the west side of East Porth on Tean.

The Bronze Age is synonymous with metal working, but very few metal objects of this era have ever been found in Scilly. Some unidentified bronze objects were found on Nornour, and, most spectacular, an Irish gold bracelet of 1000 BC was found on a beach on St Martin's. Interestingly, the ancient Phoenicians, who were great craftsmen of metal and who ruled over the eastern Mediterranean from 1000 BC–600 BC, were said to have found tin on islands to the west of Britain which they called the Cassiterides (or Tin Islands). No signs of tin mines have ever been found on Scilly but their remains could have been submerged by the rising sea. Certainly veins of tin run from Cornwall under the sea and outcrops would have surfaced in the islands. These could have soon been exhausted and then the Cornish tin could have been exploited. On the other hand, perhaps the Phoenician traders used Scilly as a port of call while on voyage to Cornwall for tin.

Around 500 BC new invaders from the Continent came to Britain. These Celts, as they were called, brought with them a new material, black in colour: iron. These Iron Age Celts were artistic, talented, aggressive and warlike, and they built massive hilltop forts all along the coast of southern England. But the Celts seemed to have made little impact on the islands except that it was around this time that the earliest fortifications were being constructed in Scilly. These cliff castles, as they were called, were coastal hill forts, possibly serving as economic and social centres under the control of tribal chiefs. Two cliff castles have been identified in Scilly, Giant's Castle on the south-east side of St Mary's, and the other on Shipman Head, Bryher. A third possibly existed on Burnt Hill, St Martin's. Interestingly, these three sites are equidistant from each other on what would then have been one large island and could have represented the tribal divisions of Scillonian society at that time.

It is important to realise that each of these different ages did not suddenly begin when one ended, but each fused and overlapped one with another. The Iron Age fused with the Bronze Age, just as the Bronze Age did with the Neolithic, and the Neolithic with the Mesolithic.

The Roman Occupation (AD 43–410)

The Isles of Scilly appeared to be little affected by the Roman invasion of Britain in AD 43. The Romans did not come to exterminate or enslave the native inhabitants of Britain, but came to subdue and rule over them and acquire another province for the

4

Roman Empire. In this they were successful. Classical Roman writers of the time mentioned Scilly as Insula Sillina and there is no evidence that the islands themselves were ever formally conquered or under the direct rule of Rome. More likely, Scilly and its people were left to look after and administer themselves. The islands were remote and it is possible that a penal settlement was kept in Scilly, since it is known that two heretic Spanish bishops were exiled there in about AD 384. Some confirmation of Roman influence in Scilly has been revealed. A classical Roman altar was discovered on St Mary's, a Roman bronze brooch was found on Gugh and a few Roman coins were found on Samson. In 1962 after a series of south-eastern gales, a number of stones placed one above another were uncovered on Nornour in the Eastern Islands and subsequent investigation revealed numerous Roman artefacts, none dating beyond the fourth century. A detailed excavation was made of the site and several small huts built of drystone walling extending along the northern edge of the island were exposed. Because of the large quantity of Roman objects found, it is conceivable that for as long as 300 years this site was possibly a shrine to a marine goddess attracting offerings from travellers between Gaul and western Britain.

During the Roman period the domestic economy of the islands did not change radically. Dogs were recorded for the first time and the red deer may have died out due to over-hunting and the reduction of woodland. A few new species of fish have been identified, including mullet and John Dory. Birds of brackish and fresh water such as teal, snipe and heron were more common, and this may indicate the rising of the sea level and the creation of pools behind sand dunes breached during storms.

The Romans left Britain in the fifth century when they withdrew back to Italy to defend their homeland against barbarians invading from eastern Europe. The survivors of their civilisation, the Romano-British, were left to fend for themselves against new invaders from the Continent.

Early Medieval Period (AD 410–1066)

The daily lives of the people of Scilly would have changed little with the departure of the Romans. Around AD 450 the seas were rising as the glaciers melted and the low-lying countries of northern Europe were reducing in size. Thus, there was good reason for their Saxon people, hungry for land, to look to a relatively sparsely inhabited England. Little is known of Scilly around this time and it appears to have been undisturbed by the subsequent Anglo-Saxon raiders of England. There is a story that King Athelston of Mercia sailed with a fleet from near Plymouth and subdued Scilly in AD 927 and subsequently established the monastic cell of St Nicholas at Tresco. Later in that century King Olaf of Norway was said to have been converted to Christianity by a holy man, Elidius of St Helen's.

There is no doubt that the islands were a convenient point for sailors and merchants plying their trade between the Mediterranean and Britain, to land for fresh water and provisions. Inevitably Christianity came to Scilly and the first stone chapels were built in the sixth century. These were rectangular in shape and three still survive as ruins today on Tean, St Helen's and St Martin's.

The Medieval Period (1066–1540)

By the time of the Norman Conquest the submergence of the main land mass of Scilly into separate smaller islands was all but complete, and one can only imagine the anguish of the islanders in those centuries long ago as their dwellings and arable land was eroded by the advancing sea and sand.

After the Norman Conquest The Islands of Scilly became the property of the Crown of England. From the twelfth century the administration of Scilly was split into two; the Abbey of Tavistock took over the northern half, including the Priory of St Nicholas on Tresco, where its ruined church is still visible in the gardens today, and Richard de Wika of St Mary in north Cornwall took over the southern half. This southern half, which is now St Mary's and St Agnes, was later transferred to the Blanchminsters, also of north Cornwall.

The Normans were the last successful invaders of England, and one would have thought that their influence would have stabilised and cemented relations with the Continent, but too often in the centuries that followed, England was at war with France, Germany, Holland and Spain. The Hundred Years War of the fourteenth century was a particularly prolonged period of fighting, when the French throne was claimed by Edward III. Both sides made continuous raids on each other. The islands were poorly defended although Ranulph de Blanchminster was given proprietorship of the natural harbour on St Mary's Porthenor (now Old Town) early in the fourteenth century by the King, in return for maintaining 12 men-at-arms for keeping the peace. Despite this, the monks would have been sorely harassed by raiders from the sea and it was probably for this reason that even before the Reformation many of Scilly's places of worship, such as those on Tresco and St Helen's, fell into decline about this time.

By the sixteenth century, because of the difficulties in defending the islands from invaders in terms of money and manpower, many of the monks were keen to leave to seek posts as parish priests on the mainland, where life would be easier. The dissolution of the monasteries in 1539 by Henry VIII hastened the decline.

In 1547 the Lord Admiral, Sir Thomas Seymour, bought Scilly from the Crown. He had been married to Catherine Parr, the sixth wife and widow of Henry VIII. Once Catherine Parr had died he plotted to marry the young Princess Elizabeth. His ownership of Scilly was marred because of his deals with marauders and raiders, and for allowing the islands to become a base for pirates. For this and other matters he was beheaded in 1549 and the islands reverted to the Crown.

With the accession of Queen Elizabeth I to the throne of England in 1558, a new era of international relations began which was to change for ever the nature and importance of Scilly which up to then was an impoverished backwater but would now become more important as a strategic military base in the Western Approaches.

Post Medieval (1540–Present)

In 1570 Queen Elizabeth gave a 38-year lease of the islands to Thomas Godolphin of Cornwall for £10 per year on condition that he defended them. Subsequent leases to the Godolphin family were to last almost continuously for 281 years until the link was severed in 1831 when Scilly returned to the direct control of the Crown and the stewardship of the Duchy of Cornwall. In 1834, an energetic Victorian named Augustus Smith took over the lease of Scilly. It was he who did so much to improve the lot of the islanders by adopting an autocratic rule in reorganising farm lands into larger plots through a revised system of inheritance, where land passed to the eldest son, and the other offspring were forced to find alternative employment. He financed and expanded existing industries, and developed new local ones, notably that of flower growing; he built schools on all the main islands and he made education compulsory. He had a new residence built for himself next to the ruins of the medieval priory on Tresco, which, when he arrived in 1834, was merely a bare windswept island without any trees. When he died in 1872 he left his successor a unique 7-acre subtropical garden with many species of trees and plants. He died a bachelor and his successor was his nephew Thomas Algernon Dorrien Smith. The same family continued the good work in increasing the prosperity of the people, and to this day the direct descendant, Mr Robert Dorrien-Smith, still leases the island of Tresco from the Duchy of Cornwall. He was the first Chairman of The Isles of Scilly Environmental Trust, which was established in 1985 to conserve and protect the environment of Scilly and to educate people about the history and wildlife of the islands.

For almost 400 years after Queen Elizabeth a complex series of defences in the shape of castles, forts and blockhouses was built on the islands, many of them still in a good state of repair. These defences reflect much of Britain's history over the years from the Spanish Armada in 1588, through the Civil War (1643–51), when the islands were unsuccessfully defended for the King, to World Wars I and II, when they accommodated hundreds of men from all three services.

Throughout this post-medieval period everyday life in Scilly was never easy and sometimes could be very hard and austere. After the Napoleonic Wars of the eighteenth century, the deprivation and difficulties of the islanders were so widespread that a committee was set up on the mainland to help them. Nine thousand pounds (a large sum in those days) was raised and most of this was used to start a mackerel and pilchard fishery. However, it was not until the arrival of Augustus Smith in 1834 that economic revival and some stability were established.

Farming has always been an important factor in the economy of Scilly. Pigs and cattle were reared and the main vegetable crops were potatoes, barley and wheat. In the nineteenth century new varieties of potato were introduced which cropped earlier and these were used as a valuable export. The farmer is blessed with a naturally fertile soil and a favourable climate, but the lack of good hedges in early times meant damage to crops from severe winds and salt spray.

Fishing and farming go together on Scilly, as in its simple industrial structure there was no room for specialisation and the same individual practised both these vital activities. There is no lack of varieties of fish and shellfish around the islands, and since the very beginning they provided a rich and important source of food, with ling, cod, pollack, plaice, crabs and lobsters particularly plentiful. Cured fish, laid on top of stone hedges or hung on the walls of houses, once constituted a staple part of the Scillonians' winter food. A poor spring and summer season of fishing meant hardship in the winter.

Early in the eighteenth century two important industries were established, piloting and kelp-making. An increase in seaborne trade and the expert seamanship of the islanders made piloting a logical and practicable pursuit. Piloting of vessels through the hazardous waters around the islands had always been practised in a limited way but there was no real organisation to control and administer it, so then pilots were self-appointed and their numbers varied. They would meet incoming vessels and would not only guide them through safely to port, but would also provide services to the home-coming sailors, like exchanging fresh eggs, meat, fish and home-brewed beer for the delicacies and luxuries of foreign lands such as tobacco, tea, sugar, fruit and wine.

Scillonian pilots used long rowing boats called 'gigs' for their work and in the last century each of the inhabited islands possessed several gigs. Then it was a race to put a pilot aboard a vessel requiring assistance and now 'gig-racing' is a recognised and important sport. Today, there are only two pilots who are appointed by the Duchy of Cornwall by recommendation of Trinity House Pilots Association.

The islands of Scilly form a 10-mile wide barrier at the entrance to the English Channel and for centuries they were unlit and basically uncharted. Shipwrecks were numerous, despite the daring, courage and expert seamanship of Scillonian pilots who brought many a doomed vessel through gales and raging seas to safety. Wrecks provided a useful addition to the harsh and sometimes poverty-stricken lives of the islanders in terms of timber, rope, tinned food, flour, soap and tea. Money, too, from wrecks was very important. In the late nineteenth century a crew of a pilot gig could expect 5 shillings each for bringing news of a wreck to the Lloyds agent and considerable sums were often paid for rescue and provision of survivors. For instance in 1910 after the SS *Minnehaha*, bound for London from New York, struck rocks in dense fog carrying general cargo including 230 head of cattle, £5 was paid for every cow landed alive.

The kelp-making industry originated in Scilly in 1684 when a man named James Nance came from Falmouth and settled on the island of Tean for the sole purpose of making kelp. He had learnt the trade in Cornwall and seemed determined to introduce the skill to Scilly. He and his family built a house on Tean, the ruins of which can still be seen. Kelp is the ash produced by burning certain types of seaweed common on the islands, such as *bladder wrack* and *driftweed*. It was used for the manufacture of soap, bleach and glass and was principally exported to the manufacturing towns of Bristol and Gloucester on the mainland. A great industry developed on Scilly employing many people. The seaweed was collected from the sea and the seashore and then placed on the ground to dry. After drying it was placed in specially made pits, 2 to 3 feet deep and about 4 feet across and lined with stones. There were about 130 of these so-called kelp pits in Scilly and most were located on a grass verge above the shoreline. The seaweed was burnt between June and August and it was reputed to have been an impressive spectacle to see the white smoke rising in the air from over 100 pits scattered all around the islands. The smell, however, was both offensive and penetrating, and in high summer windows and doors of the houses were kept tightly closed. It took about 20 tons of seaweed to produce 1 ton of soda ash, so it was a highly labour-intensive industry. In 1684 1 ton of ash could sell for 20 shillings a ton, rising to 44 shillings a ton in 1794. After the Napoleonic Wars of 1799–1815 the kelp industry fell into decline when new chemical processes for making the by-products of soda ash were found, and its price fell rapidly to only some 12 shillings per ton. The kelp pits ceased to be used in 1835 as the last flames of this once great industry were put out.

As the kelp industry withered so another grew up. Shipbuilding started in 1774 with the construction of vessels to carry kelp to Bristol. By 1850, with the help and leadership of Augustus Smith, the industry was booming, with ten shipbuilding yards on St Mary's. Most of the ships were small, under 300 tons, and built for uses such as the transport of fruit from the Mediterranean. The development of bigger iron-built ships, propelled by steam in place of wind, signalled the end of the industry in Scilly and by the 1880s all the boatbuilding yards had sadly closed.

The development of steam also brought the decline in the numbers and importance of the Scilly's pilot cutters and gigs, as ships no longer had to wait amongst the islands for favourable winds to propel them on their journey.

As we have already seen, Augustus Smith had established a wonderful garden at his home on Tresco, but it was his successor T.A. Dorrien Smith who did so much to establish a flower industry in Scilly. He studied the Dutch system of cultivation and introduced many new kinds of bulbs. In 1881 he purchased privately 44,000 bulbs in 60 varieties and by 1886 he had added another 761 varieties. He had built steam-heated glasshouses and potting sheds, and in 1889 he extended the quay at St Mary's to its present length, to cope with the shipping of flowers. The flower industry flourished and spread to all the inhabited islands. By 1931 some

212,200 boxes of flowers were exported, mainly to mainland Britain. The trade suffered badly during World War II, when export restrictions were brought in and many fields were turned over to growing vegetables. By 1989 only some 80,000 boxes of flowers were exported but the industry still gives employment and a living to many islanders. The great grandson of T.A. Dorrien Smith, Robert Dorrien-Smith, still owns the lease on Tresco and continues the family's interest in horticulture and the flower industry.

This has been but a brief history of The Isles of Scilly, of its people and its industries. The islands and their hardy and friendly inhabitants have a unique place in the wider context of the British Isles and its past influence on world events. It has not been possible to describe every facet of life on Scilly in a book like this and much, much more could have been told about the exploits of its people, of shipwrecks, smuggling and lighthouses. However, I hope I have set the scene for the description to follow of individual places, to touch again on their history and to present a picture of their birds, butterflies, plants and other wildlife.

The ocean dominates, of course, and as we shall see it has a profound effect on the natural history of the islands. In winter, sit anywhere on the miles of coastline and see the dark green walls of water pounding on the rocky shore and the spumes of spray flying across the beaches; and in summer, see how the colours change to deep blue with waves a sparkling white. Sit too on the golden sands of Tresco and look eastwards across the shallows and listen to the gentle silence of the sea here amongst the lovely Isles of Scilly.

Fulmar Petrel

11

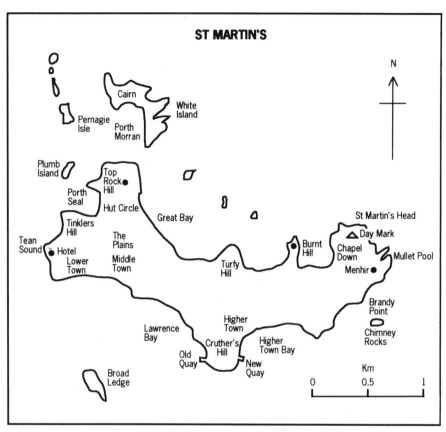

ST MARTIN'S

N

Cairn

White
Island

Pernagie
Isle

Porth
Morran

Plumb
Island

Top
Rock
Hill

Porth
Seal

Hut Circle

Great Bay

St Martin's Head

Day Mark

Tinklers
Hill

The
Plains

Burnt
Hill

Chapel
Down

Mullet Pool

Tean
Sound

Hotel

Lower
Town

Middle
Town

Turfy
Hill

Menhir

Brandy
Point

Lawrence
Bay

Higher
Town

Cruther's
Hill

Higher
Town Bay

Chimney
Rocks

Old
Quay

New
Quay

Broad
Ledge

Km

0   0.5   1

The Bishop – St Martin's

Menhir – St Martin's

2

ST MARTIN'S ISLAND

Tinkler's Rock, St Martin's

The boat journey from St Mary's quay to St Martin's takes about half an hour. It is mid-morning in June and the sun shines down from a clear blue sky. The sea is calm, the tide is low and as the boat glides over the surface one can see the beautiful colour of the water, pale turquoise with many shades of green as the sunlight penetrates to the sandy bottom. Suddenly dark shadows appear, lurking mysteriously beneath the surface; are they the remains of ancient Roman walls and ruins submerged by the rising sea thousands of years ago, or are they, more probably, beds of *common eelgrass*, or perhaps *thong weed*?

Soon the great white sweep of Lawrence's Bay appears and our destination, the stone quay of St Martin's on the Isle Hotel, comes into view. We disembark and look back at the sandy beaches of Tresco across the restless sea where a slight breeze has caused small waves to dance on the surface with little specks of white reflected from the sun's rays.

The hotel, one of the finest on the islands with an excellent restaurant and sea view, is the place I choose to stay. St Martin's is full of history and scenic beauty, and is the subject of my first chapter concerning the natural history of Scilly. The island is approximately 3 kilometres long by 1 kilometre wide and is surrounded by white sandy beaches and many offshore uninhabited rocks and islands that make a safe haven for a wide variety of plants and birds.

The history of St Martin's is interesting; Chapel Down, the easternmost part of the island, is rich in archaeological remains. A prehistoric field system covers the area, its boundaries marked with low turf-covered banks of stone which in early summer show up yellow from the flowers of *common bird's-foot trefoil* growing upon them. During the 1940s a carved piece of granite in the shape of a human head was found in a field wall and this is thought to be of a statue menhir or Celtic idol some 2000–3,500 years old, possibly the oldest carved statue in Britain. It has been reset on bedrock near a Bronze Age entrance grave with side slabs but no capstones. Prominent on Chapel Down is the red and white conical-shaped granite Day Mark which was erected in 1683 as an aid to shipping. Just beside it, and barely visible, are the foundations of an eighth-to-tenth century chapel.

Many more historic sites exist on the island. Cruther's Hill just south of Higher Town has many Bronze Age graves; Top Rock Hill north of Lower Town has a field system, a prehistoric round house and more graves; and at Burnt Hill on the north coast there are two more round houses, a field system and a cliff castle or possibly an unenclosed ancient settlement.

What an island for birds. Come with me on a day in May or June to those little coves of Norrard Bight and Southward Bight just to the north of Mullet Pool on the eastern side of the island. Gulls of many types call raucously as they busy themselves building nests on the steep rocky slopes. Keats aptly described them in his poem 'Epistles':

> The broad-winged seagull, never at rest
> For when no more he spreads his feathers free
> His crest is dancing on the restless sea.

But what is that bird that seems to delight in passing so close by, rising and falling with the air currents and giving you a friendly but curious look? This is the *fulmar petrel*. About 40 centimetres long, the fulmar is creamy-white with a smoky-grey back and tail and a very distinctive thick, short yellow bill with 'tubed' nostrils. The fulmar is an oceanic bird that spends much of the year wandering the deep Atlantic, returning early in the year to land and to breed. The spread of the fulmar around the coasts of Britain is an interesting one. At the beginning of this century the bird was known only to breed in the Shetlands and on islands off the Outer Hebrides (especially St Kilda). Now the fulmar is breeding around most of our coasts, sometimes in large numbers. The reason for this remarkable spread is not clearly understood, but is thought to be connected with changes in fishing practices and the spread of trawlers. Fulmars were first seen in Scilly in 1937 around Men-a-vaur but were not known to breed until 1958. Some 20 pairs of fulmars now nest regularly on St Martin's on low cliff ledges. There is no nest as such, the single white egg being laid on bare soil or rock with occasionally a few pieces of dried grass or sea thrift being added. The fulmar has a good defence mechanism to ward off human predators, rats or other birds that might be tempted to interfere with its egg or young. It will spit out a vile green stomach oil with deadly accuracy for up to 3 metres, which can temporarily blind or cause irritation of the skin. But of course if you do not disturb its nest, the bird is a delightful friend to watch. Sit down on the warm granite rocks and watch the fulmars as they entertain and show off, gliding, banking and turning, seemingly dancing on the wind that blows over the edges of the low sea cliffs. *Kittiwakes*, those pretty little gulls of the wide ocean, also nested on these cliffs in the early 1990s, but as I write they have disappeared from here and have moved to Tresco, as we shall see later. Other interesting birds that nest nearby are the *Manx shearwater*, the *storm petrel* and possibly a pair of *kestrels*, but more of these in subsequent chapters.

Walk over The Plains, where the alluring coconut smell of the yellow gorse mixes with the sharp tang of seaweed on the white secluded shore of Great Bay. Then you hear it, a persistent scolding note, 'wheet, tsack, tsack,' like two stones being knocked together. Look in the direction of the noise and there you see it, a male *stonechat* with its distinct black head and throat sitting on the top of a blackberry bush. This bird is common all around the islands and builds its nest on or near the ground amongst grass and brambles or at the foot of a gorse bush. The nest, made of roots, grass and moss with a lining of hair, feathers and fine grass, is hard to find amongst the dense growth characteristic of its habitat. Four to six eggs, pale bluish-green mottled with reddish brown spots, are laid in April or May.

A related species to the stonechat is the *wheatear*, which I have often seen in May and June amongst the low stone walls just above the causeway to White Island. The wheatear is a migrant species arriving in the very early spring from central Africa. It brings with it the anticipation of warm sunny days, as it flits across the fields long since abandoned; and flying from rock to rock displays its conspicuous white rump and black tail. The male has a blue-grey back in the breeding season, which turns brownish in the autumn. The wheatear is said to be an infrequent breeder on the islands but I have watched a pair showing all the characteristics of having a nest amongst the stones of the old dry wall below Top Rock Carn overlooking White Island. Both birds looked agitated, their heads bobbing up and down as they hopped nervously from perch to perch. Their nest is made of grass and moss with a lining of rabbit's fur, hair, feathers or wool, and their first eggs are laid in April with second clutches in June. Normally there are four to seven eggs in a clutch and they are a pale greenish-blue and unspotted. I did not search for the nest on that warm day in June for fear of disturbing the birds unnecessarily but I am certain they were breeding. The wheatear used to be a common bird on the South Downs of Sussex but it has sadly declined greatly since the nineteenth century, when shepherds discovered that they could catch the birds fairly easily with nets and sell them as a delicacy to the markets in London, where they made two old pence a brace. Sometimes shepherds caught as many as 80 dozen wheatears a day – no wonder they are now scarce! I wonder if the ancient peoples of Scilly caught and ate migrating wheatears as the birds made their way across the sea to the mainland. Perhaps the bird was a common breeding species here many years ago.

The tide is falling inexorably and the boulder-strewn causeway to White Island is gradually being exposed. As you walk across the slippery strands of *bladder wrack* and *thong weed* covering the rocks, be aware that in approximately four hours' time the tide will be in the same place but rising, so do not get trapped on the island. White Island is a wonderful place; here on your own uninhabited isle you can sit for hours looking out at the blue Atlantic ocean, and the great dome of sky stretched over far horizons. Peace and tranquillity are all around, and the silence is only broken by the call of birds and the gentle noise of waves rhythmically falling onto sand and rocks. Suddenly, your eyes are drawn to a large speckled cream head that appears above the surface of the sea a few yards offshore: the head of an *Atlantic grey seal* that is so common around the islands; watch it as it submerges only to reappear a few minutes later in a different place. Another and yet another one appears, until a whole group of seals are watching you. The headland at West Withan is a magical place in early May when *sea thrift* covers the rocks with a carpet of pink in which *black-backed gulls* and *oystercatchers* have their nests. In places, patches of *bird's-foot trefoil* and *English stonecrop* add touches of yellow, pink and white to the kaleidoscope of colours. Descending from the ancient entrance grave on top of the headland, look down at Chad Girt, that deep cleft in the rocks that threatens to cut the island in two at the next storm and high tide.

As you retrace your steps, look for the old kelp pit beside the path on the west side of the island. Sit beside it for a moment on the warm dry grass and let your mind wander back some 300 years and contemplate the scene. Dried seaweed lies all around and soon this is gathered up by hardy Scillonians and placed upon the pile of dry gorse and bracken that is already burning brightly in the pit. The suffocating smell of burning seaweed is obnoxious and the kelp makers keep well upwind of the fire. Tall columns of white smoke can be seen rising into the sky from distant Tean and Tresco as well as nearby St Martin's. What a sight it must have been, this ancient industry of long ago that provided a livelihood for so many people.

Look across the white sand of Porth Morran, slowly disappearing with the fast-rising tide, to the steep tall rock of Plumb Island. Not so long ago in the early 1990s several pairs of *common terns* nested on this boulder-strewn islet amongst the cast-off debris of man; ropes and netting of all kinds, bottles, cans and assorted pieces of plastic. Their two or three buff-coloured eggs boldly marked with light and dark brown spots merge so well with their surroundings that they are very difficult to spot. No actual nest material is made although a few pieces of dry grass and thrift are sometimes used as a lining. These lovely grey and white birds with a red bill and black cap arrive in May, having flown some 6,000 miles from southern Africa. No wonder they are also known as 'sea-swallows'. For some reason the birds have moved away from Plumb Island, but they can be seen everywhere around the islands flying with graceful wing-beat over the sea and then suddenly hovering on their long slender wings before diving headlong into the water emerging with a *sand eel* caught firmly in their bill.

As you scramble back onto the main island, look up at Top Rock Carn; what does it remind you of? Before long you will recognise it as the shape of a bishop with a mitre on his head. That's why this particular landmark is known as 'The Bishop'.

A bird that you may see anywhere in the area is the *raven*. It is a bird that has only recently begun to breed in Scilly. Since 1981 this, the largest of the crow family, some 60 centimetres long, used to breed on Men-a-vaur but has left there and is now thought to nest on the eastern cliffs on St Martin's. The raven is unmistakable, with its very long slightly hooked beak, heavy build and colour of intense black. It has a deep harsh croak which some construe as an evil sound that carries hidden mockery, and indeed the bird has long been a subject of superstitions, awe and legend, often being regarded as an omen of death and disaster. This reputation may be in some part due not only to its black colour, but also to its habit of feeding on dead carcases. It is indeed one of nature's great scavengers and carrion-feeders and it will eat just about anything, from seashore refuse to small animals, worms and frogs.

The raven's nest is inevitably built here on a rocky ledge and consists of heather stalks and dry seaweed, with an inner lining of grass and hair. The five to seven greyish-green eggs, blotched and spotted with brown markings, are laid between February and April. Ravens mate for life and only death will cause the pair to seek a new spouse. Watch the parent birds as they seek food of any kind to feed their hungry young. Sit on a thrift-covered ledge and see the cock raven patrol its territory, wheeling and diving, a speck of fierce black set against the deep blue of sea. Suddenly it rolls over on its back in mid-air, gliding upside down for an appreciable moment before righting itself. These joyous movements and habits of the bird seem to flow to one's own person and you feel uplifted by the sight. Let us hope this majestic bird will long remain and even spread on the remote islands of Scilly.

Wherever you go on the island you will almost certainly come across *rabbits* and indeed you will see them on all the larger islands and some of the smaller ones as well, where black ones are not uncommon. So it is worthwhile writing a few words about them. Rabbits were introduced to Scilly as a source of food in the Middle Ages and with few natural enemies – foxes, weasels and stoats are absent and the buzzard is only an uncommon visitor – they have flourished and multiplied. Only the onset of myxomatosis in the 1950s and 1960s halted their spread. Rabbits are social animals and live in colonies, where they construct a maze of burrows to form a warren. They can live just about anywhere and their burrows can be found near the top of hills and amongst the sand dunes beside the sea. As we shall see, their empty burrows are often taken over by birds such as *puffins*, *storm petrels* and *shearwaters* and used as nesting sites. Rabbits breed throughout the year, but mainly between January and June. The female rabbit, or doe, as she is known, digs a separate burrow, which she lines with grass and fur. She can produce a litter of between three and six young every month but in reality this seldom occurs and on average a doe will produce ten live young per year. The young stay below ground for a week or so, being suckled daily by their mother; she blocks off the entrance when she leaves, to protect it against enemies and to conserve heat. After about a month the young rabbits are able to fend for themselves and after about four months they are able to breed. Rabbits can be seen more often at dawn and dusk, when they emerge from their burrows to feed on grass and leaves. They will each eat up to 500 grammes of fresh green food a day and can do considerable damage to the plant ecology of an area. There is little doubt that too many rabbits in one area is harmful to vegetation and wild flowers, and in many places they maintain a grass sward so shallow that only dwarf plants can survive. On the other hand they do help to control the spread of brambles and bracken and they keep the soil well ferti- lised.

In the sandy soil of The Plains above Middle Town look beside the main path and see if you can find the *upright St John's wort* with its slender, round, smooth stem and five-petalled yellow flowers tinged with red and edged with black glands. It is a rare plant on the islands and is found in only a few places where the land has never been cultivated. Also known as *beautiful St John's wort*, it flowers from May to July. Another St John's wort species, the *trailing St John's wort*, is more common in Scilly and can be found trailing along walls, on heathland and beside roadside gutters. Like other flowers with golden-yellow sun-shaped discs, the St John's worts were once associated with the old Norse god, Baldur. In later years, through the teaching of the church, Baldur was replaced by St John the Baptist, whose festival in midsummer is marked by many remnants of sun worship. The flowers of these plants resemble the sun's rays and they were therefore regarded as a specially powerful antidote to the evil practices of witchcraft.

Nearby in the grass look out for the inconspicuous *common bird's-foot* which has small white pea-like flowers and trails through the grass. With our hand lens look closely at the flowers and see the red veins running through them. The seed pods are grouped together at the ends of stems in claw-like bunches which give the plant its name. Do not confuse this flower with the *common bird's-foot trefoil* which is a different species growing proli- fically everywhere, with its much larger bright yellow flowers.

In your wanderings over The Plains look out, in early April, for the *small adder's-tongue fern* in the shape of a single very unfern- like frond which divides to form an oval 'leaf' enclosing a spike in the shape of a tongue which carries the spores. The spores germi- nate and form male and female organs which fuse with the aid of moisture and develop into new young shoots the following spring. This species likes the well-drained maritime turf that exists on the heathy and sandy soils to the north and west of Middle Town. The small adder's tongue dies down in midsummer and nothing can be seen until the shoot reappears early the following year. This fern is a scarce plant on the mainland of Britain but here in Scilly it is fairly common on most of the inhabited islands.

A widespread plant found on the mainland but very rare in Scilly is the *common rest-harrow*. This likes calcareous soils near the sea and seems to favour those found on The Plains of St Martin's, so look from May to July for its pink pea-like flowers in the grasses beside the paths where outcrops of blown sand are found. It has a tough underground root system which used to delay the passage of harrows or horse-drawn ploughs on the mainland; hence its name. In ancient times the young leaves of rest-harrow were preserved in vinegar and used as a pleasant sauce to be eaten with meat. A liquid extract of its flowers and roots were also used by herbalists to treat disorders of the urinary tract.

As we take the path down to Middle Town, on either side you will see the newly planted coniferous *pines* of many species designed to create woodlands for the future. Perhaps when these mature, and can give wind protection to a more mixed woodland environment, then birds that once used to breed in Scilly such as the *nightjar* and *bullfinch* might return once again to nest there. Protection from the salt-laden winds is essential for the growing of crops, and as you look down on the fertile south side of the island you will see the cultivated field strips surrounded by tall hedges which act as windbreaks. One of the most efficient shrubs used as a windbreak is the *pittosporum*, which comes from New Zealand. The leaves are dark green with downy white undersides and the wine-coloured flowers which appear in April are followed by walnut-sized seed pods. These split to reveal numerous black seeds, which are eaten by birds, and hence the shrub has spread and is now found growing in many unexpected places such as on beaches and on the top of boulder-strewn hills. Some of the best shelter is provided by a native of Japan, *euonymus*, which was introduced in about 1860. It has bright green leaves, which are highly nutritious and are often fed to cattle. It has small greenish-white flowers and pinkish berries. Other windbreak shrubs such as *olearia* from the Chatham Islands in the Pacific, and *escallonia* from Chile were introduced by the Dorrien-Smith family to the Abbey Gardens on Tresco and have now spread widely on The Isles of Scilly.

Another alien plant introduced from the Pacific regions is the

New Zealand flax. This has spread to most of the islands from Tresco, where it was established in the Abbey Gardens in the middle of the last century. A tall plant, with over 1-metre long narrow pale-green pointed leaves, it has a flower spike carrying groups of green and purplish flowers which are full of nectar and well visited by insects. This flax is an important economic plant for the manufacture of linen and has been grown as a cash crop in the Isle of Man, Connemara in Eire and Wigtownshire in Scotland. Augustus Smith attempted to grow it commercially in Scilly but without success. You cannot fail to see it around the loose sand dunes of Great Bay.

In early May, sit beside the many bulb fields on the southern part of the Island. Here, *gladioli* now grow wild beside the hedgerows and fields, their spikes of purplish-red flowers contrasting well with the banks of white *arum lilies* that grow amongst them. The gladioli are known locally as 'whistling jacks' because it is said their seed pods 'sing' in a strong wind, but more probably children use the leaves as reeds to whistle through. A good indication of how the gladiolus has spread itself far and wide is to see clumps of it swaying with the breeze on the sandy dunes around Great Bay. In the autumn, *belladonna lilies* can be seen on the south-facing slopes, their delicate colour of rosy-pink contrasting beautifully with the blue sea and sky.

Many interesting wild flowers can be found in and around the bulb fields: *spring beauty*, *Bermuda buttercup*, *three-angled leek*, *rosy garlic*, and *English catchfly*, to name but a few. Spring beauty is easily recognised from its curious small white five-petalled flowers, which rise from the centre of rounded fleshy leaves. It is very common around the cultivated fields of Lower Town in the early spring. It is a native of North America and has established itself on sandy soils all around Britain over the past century.

The Bermuda buttercup is a common and invasive weed carpeting the soil between the rows of bulbs. It is not a true buttercup but is a member of the oxalis species and has large bright yellow flowers with many leaves forming a rosette at the base and springing from a buried bulb. This beautiful plant is an introduction from South Africa and despite its name does not originate from Bermuda. It flowers from March to June and is one of my favourite flowers of Scilly.

Three-angled leek, named after the shape of its stem, flowers abundantly around the fields, its large white flowers prominent amongst the fresh green grass of spring. This flower has spread to most of the other islands, unlike another of the same family, rosy garlic. This lovely plant flowers later in June and is only found on St Martin's and St Mary's. Look for its bunches of beautiful pink coloured flowers swaying in the breeze on top of 30–60 centimetre-high stems in the grasses beside the roadside or field edge.

English catchfly, related to the campion flower, is another common plant of the bulb fields. It is a hairy, sticky plant with spoon-shaped lower leaves. The small white or pink flowers are borne singly between the upper leaves and the stem and bloom between May and October. Occasionally you will find both white and pink flowers on the same plant, and there is a very rare variety with a red spot situated on each white petal. See if you can find this last one on your walks.

As you pass the little general store in Lower Town, look on its walls and see the uncommon white variety of the well known *ivy-leaved toadflax*. This forms a beautiful tapestry as its long slender stems, with leaves lobed like that of ivy, trail across the granite rocks of the building. Normally its flowers are blue-purple, with yellow near its mouth to show insects where to seek admittance to the nectar. At the foot of the wall beside the road you will see the *dwarf mallow*, with downy stems and small roundish leaves that lie along the ground; the flowers about 2 centimetres across are pale lilac, with darker lines converging to the centre. This is just one of several mallow species that give colour to the islands and others will be mentioned later in this book. Both the ivy-leaved toadflax and the dwarf mallow flower from May to September.

In the fields near Higher Town look for the vivid lavender-blue dandelion-shaped flowers of *chicory* some 5 centimetres across, growing in clusters on 1-metre-tall stems. The flowers whose colour is quite unlike that of any other wild British plant, bloom from June to October. The fleshy root stock, when roasted and ground up, is used for making coffee which is popular on the Continent. For this reason it is cultivated on a large scale in Belgium and Guernsey. It is said to be very rare in Scilly but I have nearly always managed to find it on St Martin's.

As the summer flowers fade and die and the first chill winds of autumn begin to make themselves felt, the colours on the higher exposed hills change to a deep purple as the dominant plant *ling* comes into full bloom. Here on these wind-exposed downs where the soil is thin over granite rock, the ling is only some 15 centimetres high, entangled and gnarled, and eroded by the gales that sweep over it. Under the prevailing conditions the ling forms ridges whose western side consists of the exposed root systems with the flowering shoots lying prostrate directed towards the east. This maritime, or 'waved' heath as it is called, is a characteristic habitat on the islands, and here on St Martin's there are fine examples on Chapel Down and on Top Rock Hill. Good examples of waved heath can also be seen at Castle Down on Tresco, and Shipman Head Down on Bryher. The Latin name for ling is *calluna*, which is derived from the Greek word *kallino*, meaning to beautify. Those who have seen moorlands in August and September covered with this plant will agree with this apt name. In reality, however, the name was given because this heather was made into brooms which did the beautifying by sweeping. In Scilly calluna was important to its people because under conditions of slow growth over long periods of time it formed peat. This was cut in the past and was the main source of fuel and such names as Turfy Hill on St Martin's north coast commemorate the allotments where it was dug.

Amongst the lilac coloured heathland the small flowers of *tormentil* show up as specks of bright yellow. The whole plant grows low amongst the stunted ling; its lower leaves have stalks and sprout from a woody rootstock, but the deeply cut leaves on the stem are unstalked. The roots of tormentil contain tannin and are used in modern herbal medicine to treat the 'torments' of diarrhoea and sore throats, hence its English name. In former times the roots boiled in milk were used to treat stomach upsets in children and calves; the plant was also used to cure smallpox, cholera and whooping cough and as a lotion for ulcers. An important plant indeed as a 'cure-all' for many ailments.

Autumn is the time to find one of the few wild orchids that grow in Scilly, that is the *autumn lady's-tresses*. This orchid flowers in August and September, its small white flowers arranged in a very distinct spiral on a short stem densely covered in white hairs. It is only 5 to 16 centimetres tall and at first sight it appears to have no leaves. However, a closer inspection of the base of the stem reveals the withered remains of the leaves amongst the grass. Close by the stem and to one side you will find, in September, two or three very fresh bluish-green leaves which are still developing. These persist over winter and from their centre will grow next year's flowering stem. Look for this lovely little orchid on the short turf above Great Bay, on The Plains and by the shore at Lower Town.

Autumn also brings with it many kinds of fruits. The *bramble* is a study in green, purple and black and several species of this shrub grow abundantly on the roadsides, heathy dunes and cliff tops. There is a *red admiral* butterfly feeding on a blackberry with its wings folded giving it a perfect camouflage. The red admiral is one of our most beautiful butterflies when, with its wings open, it displays its red bands, white markings and sky blue patches all on a black background. Truly it is exquisite. Its presence here as elsewhere in Britain depends on migrants coming across the sea from the Continent in May and laying eggs singly on the *nettle* plant. These hatch in a week and give rise to the familiar sight of these butterflies in the summer and autumn. If you look carefully at a patch of nettles you may see the green or brown caterpillars (there are two types) feeding on the leaves from under a 'tent' made by drawing two leaf edges together with silken threads. The chrysalis with attractive gold spots can be seen suspended from nettle stems in late summer and autumn. The butterfly is territorial in nature and fiercely defends its patch of hedgerow or garden by attacking and driving away intruding butterflies or other insects. Sadly, it cannot tolerate the cold, and does not normally survive the winter, although in recent years there have been increasing reports of its hibernation in houses and outbuildings.

The *small tortoiseshell* is one butterfly that has learnt the trick of hibernation to escape the dark and cold of winter and you may find it in outhouses and even inside the house under pelmets and in curtains. Spring arrives early in Scilly, and on the first warm day the small tortoishell ventures out to join the sulphur-coloured *brimstone* as one of the season's first butterflies. The upper sides of its wings, some 5 centimetres across, are brightly speckled with orange, brown, black, yellow and blue markings, making it truly unmistakable and alluring. Strongly territorial, the small tortoishell will chase off any intruder that attempts to invade its favourite patch. Mating takes place at dusk and the butterflies often roost for the night under nettle leaves, where the tiny green eggs are laid in groups in May. The caterpillars are black with yellow stripes, and when they first hatch out, they form a mass before eventually becoming solitary in their final stage. They feed on nettles for about three weeks and remain in chrysalis form for a further 12 days. The butterflies emerge in June and July and produce a further generation that fly in August and September and then hibernate. Occasionally immigrants from France fly across the sea in July and August to supplement the numbers already present. Do watch and enjoy the spectacle of these common butterflies as they flit and skim around the red flowers of the *valerian* plant by the roadside in Lower Town and see how their colours blend so well with the blue of the sky.

At any time between May and August look out for the *common blue butterfly* darting about amidst the grasses that surround the

24

many white beaches on the island. Its caterpillar feeds on several plants that are widespread such as *clover* and *birds's-foot trefoil*, so you may see it just about anywhere that these plants are growing. Only the male of the species has the flamboyant metallic blue markings that gleam in the summer sun; the female is usually brown with orange marking around the wings. Interestingly, the bright blue colour is caused by the diffraction of sunlight by numerous corrugated scales on the wings which absorb all the colours of the spectrum except blue; therefore there is no blue pigment contained in the wings. The eggs of the blue butterfly are laid singly on the food plant and the green caterpillars, after feeding, hibernate in winter. The chrysalis is formed in spring and lasts in this stage for about two weeks. The adult butterflies emerge in April. Two generations are produced each year but the adult lives for only two or three weeks. Often you will see groups of males clustered around puddles, drinking the water for their mineral content, especially salt. As the sun goes down in the evening you may also find a number of these butterflies clinging to grass stems, resting with their heads down.

The common blue is a difficult butterfly to get close to when the sun is shining brightly, but when the sun is blotted out by cloud and in the evening when it is low with little heat, then the butterfly can be easily approached and even touched. It is clearly attracted to the warmth of one's hand, and can often be seduced to perch on one's finger.

Heat is important to butterflies as they need the sun's warmth to circulate the blood in their veins, so enabling them to fly. *Moths*, too, require heat to fly, and as most fly at night they can often be seen vibrating their wings furiously, so increasing their body temperature by muscular activity. The wing scales of moths are thicker than butterflies' as a rule and they act as a kind of insulating fur to retain body heat.

Butterflies bring great joy to people, and as one watches their beauty and form it is worth pausing to reflect on how, with their short lifespan and fragile appearance, they have evolved from the great dragonflies of the carboniferous era some 350 million years ago that had wingspans of 60 centimetres or more.

Before leaving this lovely isle of St Martin's you will not fail to notice how the tides around the coast affect many aspects of life here, from boat travel to fishing, and even collecting seaweed as a fertiliser for the garden. So, as you sit on a secluded beach and watch the inexorable changes of sea level, it is worthwhile considering how tides are formed. Two high tides and two low tides are created each day by the gravitational forces of the moon and sun. When the sun is in line with the moon on the same side as the earth, their joint gravity gives rise to extra-high tides, known as spring tides. Very low tides, or neap tides as they are called, occur when the sun acts at right angles to the moon. These spring and neap tides occur every 14–34 days. The moon, because it is nearer to the earth than the sun, has a greater gravitational pull, and because it takes 24 hours and 50 minutes to circle the earth, high tides occur 12½ hours apart and 50 minutes later each day.

Walk along the beach at Lawrence's Bay when the tide is very low and look for the lovely shells of the sea, like *dog whelks*, *painted topshells* and *blunt tellins*. Both dog whelks and painted topshells belong to the group known as univalve molluscs, which are animals with a single spirally coiled shell, or a bowl-shaped shell. The blunt tellin is a bivalve where the shell is divided into two valves, hinged together, enclosing the mollusc's body.

The dog whelk is a relatively small shell up to 3 centimetres long with up to five whirls and spiral ridges, which are flattened. Their colour varies from white, through yellow, to purple-brown. Painted topshells can be recognised by their pyramid shape and flat base and their stripes of red and purple. Both these univalves belong to animals that can usually be found living among seaweeds and around rocks, on the lower shore.

The blunt tellin is up to 6 centimetres across and has a thick smooth shell. Like many bivalves, it burrows into the sand and is hidden from view, the empty shells being brought to the surface by wave action which causes them to be broken apart.

Enjoy your shell hunting, when even on a crisp cold day in autumn or winter, you will feel a sense of calmness and beauty of places where the gentle lapping of the waves and the wild musical piping of wading birds are your only companions.

How does one leave and say goodbye to such an island where seasons of colour start in March with bluebells and yellow daffodils, and end in September with the purple of heather? Sit on the dune grasses and look down to where the white sands of Great Bay meet the crystal-clear water with shades of turquoise and deep blue, or stand amidst the ancient fields on Chapel Down and look to the Eastern Islands set in the silvery sea of a low sun in early morning.

Great Bay, St Martin's

Raven

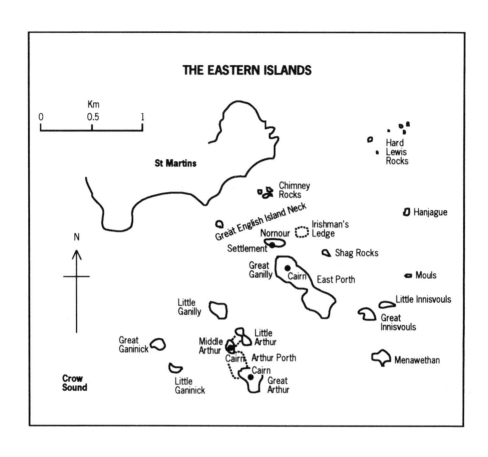

THE EASTERN ISLANDS

Km
0   0.5   1

St Martins

Hard
Lewis
Rocks

Chimney
Rocks

Hanjague

Great English Island Neck

Irishman's
Ledge

Nornour

Settlement

Shag Rocks

Great
Ganilly   Cairn   East Porth

Mouls

Little
Ganilly

Little Innisvouls

Great
Innisvouls

Great
Ganinick

Little
Arthur

Middle
Arthur

Cairn   Arthur Porth

Menawethan

Crow
Sound

Little
Ganinick

Cairn

Great
Arthur

N

Atlantic Grey Seal

28

3

THE EASTERN ISLANDS

Bronze Age Cave, Middle Arthur

As you fly into St Mary's from Penzance, your first sight of The Isles of Scilly is a group of rocks and small islets on your right hand side. These are the Eastern Islands, and with their green vegetation, grey-white rocks and little beaches of pale golden sand set in a clear blue sea, they give an enticing indication of the delights to be found on them.

Many of the Eastern Islands were inhabited when they made up just part of the greater Island of Scilly before the submergence, and well-preserved Bronze Age entrance graves are to be found on Great Arthur and Middle Arthur. The two graves on Middle Arthur have chamber sides constructed of stone slabs with slightly displaced capstones. Excavations of the southernmost grave in 1953 revealed a large funerary urn, together with some flint and bone pieces. The recently (1962–66) excavated site of an early settlement on Nornour suggested that it was possibly a shrine and place of pilgrimage in Roman times. However, before then it was probably a thriving centre of population as the well-sheltered site lies on the southern slope of the present island, above what would have been flat lands suitable for pasture and cultivation.

Sit on the rocks on top of Nornour, as I did one fine morning in midsummer, and watch little puffy white cumulus clouds build up

29

with the increasing heat of the day and then slide across the wide sky and the limitless horizon beyond low lying islands of sand and stone. Think of what it must have been like, 500, 1,000 or even 5,000 years ago. The ghosts of the hill-top graves seem almost to rise up over this beautiful place and speak about times past and how, as human inhabitants, they toiled hard to eke out a subsistence living and then died after just a short life. A breeze gets up and rustles the browning bracken against the undergrowth of bramble. The raucous calls of a group of *black-backed gulls*, disturbed by a passing boat, break your mood of timeless contemplation.

The boat reminds you of the many shipwrecks that abound in this area of shallow sea and barely submerged rocks. In 1872 the West Cornwall Railway Company's 144-ton paddle steamer, *Earl of Arran*, came aground on Nornour. The ship was carrying 92 passengers and mail for the islands from Penzance and its captain had been persuaded by a local islander to take a short cut through the Great English Island Neck to save time. This is a narrow stretch of water between Nornour and St Martin's and abounds with many low-lying reefs and ledges, invisible just below the surface at certain states of tide. The *Earl of Arran* struck rocks on Irishman's Ledge just to the east of Nornour and was only prevented from immediate sinking by being run ashore on Nornour itself. Unlike other similar incidents, no one was drowned on this occasion and today her huge iron boiler can be seen at low water lying on rocks to the west of the island; a salutary reminder of the dangers of sailing around Scilly without a good knowledge of the area and the tides.

Wherever you go in the Eastern Island group you will certainly hear and see the *greater black-backed gull*, the largest of our gulls with a length of some 70 centimetres and a wingspan of over 120 centimetres. It is easier to distinguish from other gulls, not only by its size, but by the fact that the adult is black above, rather than the slate-grey of the lesser black-backed gull. Also, its legs are whitish-pink and not yellow as in the lesser black-backed gull. It has a massive yellow bill and is fiercely predatory and is especially fond of eating the young of other seabirds. It is not an uncommon sight to see one of these great gulls alight amongst a group of puffins or colony of *terns* with newly-hatched young; seize one, shake the life out of it, rip out its entrails and quickly devour them and the rest of the bird. Such is the world of nature in the raw. The call of this big bird is harsher, stronger and more strident than that of smaller gulls. There are about 750 pairs on Scilly, nesting on cliff ledges or on the top of rock slabs and islets; the nest is made of seaweed, heather and dry grass and the two or three yellowish-brown eggs blotched with grey and dark brown are laid in May and June. The young hatch out after 27 days. When protecting its young the parent bird, like many of its species, will dive down from the sky and try to hit you on the head with its feet, at the same time calling loudly in an abusive and emphatic manner. Very unnerving to the unwary, so be careful if you are near a cliff edge or have not got your feet planted firmly on the ground.

Sit beside the cairn on the northern-most hilltop of Great Ganilly and look over to that large rock Hanjague, rising majestically out of the sea with a ring of white breaking water foaming around its granite base. A large snowy-white bird with black-tipped wings is plunging headlong into the water and raising a fountain of spray. A *gannet*; what a sight it makes as it flies about eight metres above the surface and then makes its spectacular dive for fish. An interesting point about the gannet is that not only are its bones hollow and air filled as is usual with most flying birds, but it also has an elaborate system of air sacs about the front of its body. These inflate when the bird is in flight and help to cushion the shock on hitting the water in one of its eye-catching dives. The gannet does not breed in Scilly but is seen throughout the year around the island coasts. No doubt they will have come from the nearest breeding colony on the island of Grassholm off the south-west coast of Wales.

On some of the remote and isolated rocky islets you may see one or more of the three species of auk: the *puffin, razorbill* and *guillemot*, a few of which breed here; but they are more common elsewhere in Scilly and will be described in later chapters.

One of only two resident birds of prey on Scilly is the *kestrel*, and although many visit in the autumn and stay for a month or so, only a few pairs actually nest. Easily recognised, because apart from the much larger buzzard, it is the only British bird of prey that hovers. With quivering wings and spread tail, it maintains its position for many seconds before dropping suddenly earthwards onto its intended victim, a small bird or rodent, or flying off to another vantage point. On a calm, still day in spring I remember watching a kestrel hovering over the sandy beach on Little Arthur. Suddenly a small brown rat ran across the sand and into the marram grass; the kestrel pounced, remained for a few seconds on the ground, and then flew away with the small rodent firmly caught in its claws. *Rats* are not uncommon on the islands and because they are a menace to nesting birds such as the puffin they are kept under firm control with poison. Kestrels will feed on small birds, rodents and just about any kind of insect.

The male kestrel is a beautiful bird with its bright blue-grey head, russet-red back and black-banded tail. As in most birds of prey, the female is larger and, in the case of the kestrel, has a barred rusty-coloured tail. On Scilly the birds nest on sea cliffs, where they use no nest material, but merely scratch out a small cavity in a crevice which soon becomes plentifully sprinkled with droppings. The four to seven creamy-white eggs are thickly blotched with brick-red and are laid in March and April. The female alone incubates the eggs, and after about 28 days the young hatch out; during this time the male generally brings food to the female on the nest. Sometimes he may perch nearby and call; then the female will leave the nest and receive the prey directly from him. The male's rate of hunting increases markedly after the young hatch. The newly hatched young look scrawny and vulnerable and it is hard to imagine that in less than a year they will be sleek adult hunters and masters of the sky as they patrol with fast shallow wing-beats and hover motionless on the wing, looking for prey. The young leave the nest after about 30 days and then remain in the area for about another month. They are fed by their parents until they are strong enough on the wing to fend for themselves.

Listen out for the high chattering call 'zee-zee-zee' of the *peregrine falcon*, for this magnificent bird with its scimitar-like wings breeds on the island of Great Innisvouls. Here a pair of peregrines have their nest, a mere scrape in the debris on a rocky cliff's ledge. Sometimes, however, the birds will utilise the old nest of a raven or gull. The clutch of three or four richly coloured, buff, red and grey blotched eggs is laid in March and April. Like most birds of prey both sexes take turn to incubate the eggs, the mate often bringing freshly caught food to the bird on the nest. Incubation takes about 31 days and at first the young peregrines are clad in white down, but a second coat of down is grown after 10 days. This coat of fur keeps the young falcons warm in their exposed environment when the parents are away catching food for their voracious appetite. The young fly from the nest after about 40 days and as they grow strong on the wing, they accompany their parents in search of prey which consists almost entirely of living birds. These are caught in full flight and are struck dead by a blow from the claw of the falcon, delivered with great force at the end of a near vertical dive. I remember one occasion as the sun rose over the purple-clad hills of the Pennines in England, I heard that shrill call of a peregrine and looking up, there it was flying low and fast, its head tilting from side to side searching for a victim. Suddenly it swept high into the heavens, soaring upwards, ever upwards, until it was just a speck in the pink dawn sky. I kept it in sight until the bird saw what it was looking for and then plunged vertically onto an unwary pigeon. The pigeon saw it at the last moment and tried to dodge sideways, but the agility of the falcon was the better. A small bunch of white feathers falling gently to earth marked the spot where death was instant and nature was seen in its awesome finality.

Look out for other birds of prey which are just passing migrants such as the *buzzard, red kite* and *hobby*. Others you may see such as the *sparrowhawk* and *merlin* sometimes stay for a few months in the winter.

A little bird you will see flitting around the foot of the rocky cliffs is the *rock pipit*. About 16 centimetres long, it has brown upperparts and streaked olive-buff underparts, and its dark legs distinguish it from most other pipits. The rock pipit is a common breeding bird on Scilly and is thoroughly at home where the waves pound unceasingly on the rocky foreshore. It thrives off the minute life at the water's edge and is often seen running in and out of the surf. Equally it loves the weed-draped rocks uncovered by the retreating tide. The rock pipit builds its nest of seaweed and grass lined with gull's feathers, hair and fine grasses, under an overhanging stone or in rock crevices in the shelter of a tuft of grass. The four or five greyish-green eggs are closely spotted and mottled with reddish-brown specks and are laid from April to July.

The rock pipits in Scilly are the favourite host for the *cuckoo*, which arrives after a long 6,000-kilometre flight from Africa in early April. The distinctive call of the male cuckoo heralds the arrival of spring on the mainland, although in the warm climate of Scilly spring has long started before the birds arrive. About 33 centimetres long, they have blue-grey plumage with white underparts and can be distinguished in flight by their pointed wings and long tails. The cuckoo makes no nest but looks for that of a foster bird, such as the hedge sparrow and rock pipit, in which to lay its dark grey egg. The parent cuckoo discovers its host's nest by sitting on top of a rock or bush and watching what goes on around and noting nest building birds of any species that interest her. When ready to lay, usually in late afternoon, the cuckoo glides down to her chosen nest, removes one of the host's eggs, and then lays her own. The new owner rarely refuses to accept the new egg and continues to incubate it. The cuckoo's egg has a short incubation period of just 12 days and soon after hatching out, the young cuckoo instinctively pushes all the host's eggs out of the nest, leaving itself to be reared by the foster bird. After about three weeks it leaves the nest and migrates back to Africa in September alone, using an inborn navigational skill that is quite remarkable. It is interesting to note that the cuckoo is the only British bird to foster its young in this way and by doing so is able to raise a larger number of birds than it could feed by itself. It is thought that on average a female cuckoo lays some 12 eggs each spring.

I believe the arrival of the cuckoo and its evocative calling is one of the most enchanting sounds on the islands. Relax on the sandy shore and watch and listen to the bird as it calls from the top of a nearby rock, its hawk-like form giving just the slight sense of menace to the scene. The bird flies off, no doubt to victimise another pipit. Let us leave the cuckoo and reflect on the lovely lines of Wordsworth in his 'Ode to the Cuckoo':

O blithe new-comer! I have heard,
I hear thee and rejoice.
O Cuckoo! Shall I call thee bird,
Or but a wandering voice?

While I am lying on the grass
Thy twofold shout I hear;
From hill to hill it seems to pass
At once far off, and near.

You are alone on the beach between the two hills of Great Ganilly. The sun is high in the sky, the air is warm and sultry and the hot sand burns your feet. Follow the great sense of freedom you feel and let the sea lure you into its cool embrace. Slide into its limpid waters on the lee side of the shore and endure the sudden impact of the cold water on bare limbs. But worry not, for the heat on the beach will soon bring a warm glow to your body and you will remember and want to relive the experience again and again in future years.

Nearby, around Shag Rocks and Hanjague, seals can be seen basking in the sun. There are two types of seals in Britain, the *Atlantic grey seal* which is the largest and most numerous and the *common or harbour seal*. The grey seal, Britain's largest wild mammal, can grow up to 3 metres long and about 250 of these beautiful creatures breed on the outermost of the Western Rocks and just occasionally on the Eastern Islands. In these remote and lonely places they appreciate the solitude and lack of disturbance by man. The misleadingly-called common seal is much smaller, only up to 2 metres long, and has a snub nose rather than the straighter more 'Roman' nose of the grey seal. It is found mainly on the east coast of Britain, whereas the Atlantic grey seal, as its name implies, is located around the rocky shores of western Britain.

The grey seals produce young in their breeding colonies or 'rookeries' as they are called, between September and December. About a month before the breeding season, pregnant females and bulls begin to gather at their rookeries, which in Scilly consist of just a few pairs because of lack of space. As with many mammals, the largest bulls fight or just challenge each other for supremacy. After the females have given birth, each bull establishes its territory, usually forming a harem with three to seven cows. The females mate again two or three weeks after giving birth and with a gestation period of some 350 days new pups are born almost exactly a year later. A grey seal mother generally gives birth to just one pup each year and she feeds it on land for about three weeks. After this period the young are left to fend for themselves.

After mating the seals disperse into the sea around our western coasts in search of fish or shellfish on which to feed. Seals have half as much blood again as similar land animals and therefore carry ample supplies of oxygen within their bloodstream. They can also tolerate higher levels of carbon monoxide in their blood than other creatures. These features allow seals to remain submerged under the water for periods of up to half an hour. A thick layer of blubber beneath their skin protects them from cold seas.

Seals have always been hunted for their skins and for the oil produced from their blubber. At one stage in the early 1890s the grey seal was in danger of becoming extinct but their numbers increased after new protection laws were introduced in 1914 and 1932 which prohibited their slaughter during certain close seasons of the year. Now Britain has over 40,000 breeding Atlantic grey seals, representing some 80 per cent of the world's population. As the seal population grew so did the complaints of fishermen, who were concerned about threats to their livelihood from reduced fish stocks and damage to their nets. So in 1970 the restrictions on the killing of seals were eased and young seals were allowed to be shot in certain areas, mainly in the north of Britain, where the complaints could be justified. This culling of seals may be in the interests of the animal in preventing their breeding colonies becoming overcrowded and insanitary.

The adventurous reader who wants to study the Atlantic grey seals of Scilly should contact the diving centre on St Martin's. Diving expeditions to the Eastern Islands are arranged so that you can actually swim with the seals under the clear water and amongst the brown and green kelp and other seaweeds.

The Eastern Islands are sheltered from the Atlantic gales of winter and, unlike the Western Rocks around Annet, have more soil and hence a more interesting and larger variety of plant species. Two interesting plants that grow normally as woodland species on the mainland grow on the small island of Great Gannick; namely, *butcher's broom* and *wood spurge*. Both are very local and are found only on a few of the other larger islands.

Butcher's broom is an evergreen shrub about 60 centimetres high which bears tiny greenish-white flowers from January to April, and bright red berries from October to May. The flowers appear from the middle of the dark green 'leaves', which are not truly leaves but are in fact flattened branches each with a sharp spine at the tip. It is hard to imagine that this plant is a member of the lily family. Their spiny stalks were once used by butchers to scour the tops of chopping blocks, hence their name. Look for it amongst the great boulders near the island's summit.

Wood spurge, about 60 centimetres high, is an attractive plant with evergreen leaves often tinged with red and topped by bright yellowish-green flowers. These flowers, which bloom from March to May, are held in saucer-shaped yellow bracts. Like most of the spurge family, when the stems are broken, a white milky sap exudes which is acrid in character and poisonous.

Try to imagine this group of islands many years ago when it was all part of one land mass and covered with woods of oak and birch. Perhaps the wood spurge and butcher's broom flourished in these ancient woodlands and the ones we see today are their descendents clinging on to a precarious existence.

Amongst the boulders and rocks above the sandy beach on Great Ganilly grow numerous plants of the *yellow-horned poppy*. This is one of the most beautiful and striking plants of the seashore. The bright yellow flowers, 7 to 10 centimetres across, can be seen from June to late September. Individual flowers, the lowest of which open first, last only a day or so, but are soon replaced by fresh ones growing up the stems. As each flower dies it is replaced by a narrow, curved seed pod about 20–30 centimetres long; these 'horns' are the longest of any British plant. The glaucous thick and leathery-like foliage is often untidy and sprawls over the beach. This is an unmistakable plant that is fairly common around the islands but it is a declining species in the rest of Britain and is absent from the far north.

Another plant growing amidst the shingle and stones just above the high-tide line on Great Ganilly is *sea kale*, one of the largest and most conspicuous of maritime plants. It has a long, fleshy rootstock from which arise purplish stems 30–60 centimetres high with large blue-grey heart-shaped leaves. The stout stems bear broad-domed sprays of creamy-white flowers with dark centres which bloom from May to August. When newly sprouted, the young fleshy leaves have always been considered a vegetable delicacy when cooked. For this reason it was dug up from many of our beaches and transferred to gardens and is now an uncommon plant and in decline. In Scilly it was grown extensively as a market crop in the nineteenth century but today it is found, apart from Great Ganilly, on only a few beaches on the inhabited islands.

One cannot but help notice how both the yellow-horned poppy and sea kale and many other plants growing on beaches and shingle have thick fleshy leaves and stems. The succulent nature of these plants is due to the development of large cells known as 'aqueous tissue' which are employed for storing water. In dry weather, as water is gradually lost by transpiration and by the drying effect of salt-laden winds, these water-holding cells shrink until the rains come and they expand again. Thus they act as a water-storage system for use by the plants in times of drought when their immediate environment becomes very arid indeed. These maritime plants are also equipped with long tap roots which burrow deeply into the ground in search of water. The nature of plants has certainly evolved in an impressive manner to cope with the vicissitudes of life.

One of the many species of the pink family to be found in Scilly and on most of the Eastern Islands is *sea campion*. Growing on the shoreline, in the grassy banks beside the path and adorning many of the sea cliffs, the drooping white flowers and grey-green leaves of this plant are quite distinctive. Each flower 'springs out' from a swollen bladder-like calyx with purple veins running through it. They bloom from May to August, are very fragrant and are attractive to bees and night-flying moths.

Look out for that lovely plant *golden rod*, which grows on heathy and rocky ground on Middle Arthur and Great Ganilly. Curiously it is a comparatively rare plant in Scilly and can only be found on the Eastern Islands, St Mary's and on St Martin's Head, whereas on the mainland, especially in Cornwall, it is common. From 8 to 60 centimetres tall, golden rod has lance-shaped, slightly toothed leaves arranged alternately on the stem. The upper half of the stem has numerous small branches bearing spikes of bright golden-yellow flowers which bloom from late June to September. Golden rod has been much favoured by herbalists and for hundreds of years it was used as a tincture to treat kidney infections, arthritis and eczema. Also it was used externally to sooth inflammation and encourage healing. Nicholas Culpeper, the well known seventeenth-century astrologer-physician, wrote of this plant: 'Venus rules this herb. It is a balsamic vulnerary herb, long famous against inward hurts and bruises . . . it is a sovereign herb, inferior to none.' Clearly this, like many other species, was an important plant of bygone ages.

You cannot fail to see masses of *common ragwort*, whose handsome golden yellow flowers give such pleasure to the eye in high summer. Sadly though, this plant is highly toxic to grazing animals, causing insidious and irreversible cirrhosis of the liver. On the mainland it is said to be responsible for half the cases of stock poisoning, but of course on the uninhabited islands of Scilly it does little harm. Ragwort has many other popular names, two of which, stagger-wort and stammer-wort, suggest that it was formerly regarded as a medicine for epilepsy and stammering. Common ragwort grows to almost 120 centimetres tall in places and when crushed between the fingers its grey-green, deeply toothed leaves give off a malodorous smell. Often the beautiful caterpillars of the *cinnabar moth* will be found feasting on the leaves, their alternate wings of black and orange blending in well with the colour of the flowers.

Wall pennywort, or *navelwort* as it is sometimes called, is a plant that is abundant on all the islands in Scilly. The fleshy leaves of wall pennywort are disc-shaped, with a depression in the middle suggesting a navel on their undersides, where they are attached to their stalks. Amid these curious leaves rises the flowering stem bearing pendant yellowish-green tubular flowers which bloom from May to August. In places where there is a very high salt concentration this plant thrives and grows very tall, up to 40 centimetres high, with exceptionally large leaves. Thus on Nornour and Great Innisvouls it gives an impressive appearance in rock crevices drenched with salt spray.

In your exploration of these islands you will not fail to see the lovely green fronds of the *sea spleenwort*, growing in places up to 70 centimetres long. This fern is abundant on most of the islands, where it thrives particularly well among the coastal boulders and on the sea cliffs. Today's ferns are among the most primitive plants on earth and their fossil remains date back more than 300 million years. At that time, of course, ferns were very much larger, reaching 30 metres high, and established the world's first forests. Their remains now form the basis of the present-day coal deposits.

Many butterflies can be found amongst the sheltered hollows and south-facing slopes of the larger of the Eastern Islands. Two of the rarer butterflies you could just see are the *clouded yellow* and the *monarch*.

The clouded yellow is a migrant which breeds continuously in North Africa and the Mediterranean, and every spring it migrates north. In most years this migration loses momentum in central France, but exceptionally they reach the British Isles, sometimes in very large numbers, such as in 1947, 1955, 1983, 1992 and 1997. What a thrill to see these fast-flying orange-yellow butterflies as they flit between such common plants as *clover*, *thistles* and *scabious*, seeking out their nectar. Any clouded yellows you see on the islands in spring are probably resting on their long flight from the Continent to their destination on the mainland, where they breed and lay their eggs on plants of the melilot and trefoil species. They seldom alight and spread their wings, but remain poised with wings folded ready to fly off at the slightest disturbance. Their eggs hatch out in June and give rise to a single brood of butterflies in autumn. Sadly, they cannot survive the cold, damp weather beyond November and try to migrate back across the sea to warmer lands. In this respect they are similar to that other but much commoner migrant butterfly species, the *painted lady*.

If you are very lucky you could just see the monarch, an American species of butterfly which occasionally, and just by chance, travels the 5,000 kilometres across the Atlantic, aided by favourable westerly winds. This black and orange butterfly with white spots has a wingspan of some 10 centimetres and is quite unmistakable. The caterpillar of the monarch feeds on the milkweed plant, which is unknown in Europe except as a rare garden species, and so this beautiful butterfly does not breed in Britain. In America the monarch migrates every autumn from the north of the country to as far south as Mexico, about 2,200 kilometres, and it is during this migration that it is sometimes blown off course. After their long journey across the sea they immediately seek out large garden flowers for energy-giving nectar, and in 1981 many were seen in Scilly. So keep your eyes open for them from August to October.

Having just described such a rare and beautiful butterfly, it seems fitting to bring to an end this chapter on the Eastern Islands, whose charm and captivating appeal will never let you forget them.

A good way to visit the Eastern Islands from St Martin's is to contact one of the local boatmen who will take you on an excursion to any of them in his boat during daylight hours. He will leave you alone on any deserted and uninhabited island and then pick you up at a pre-arranged time. So on a fine spring or summer day enjoy the peace and serenity which these beautiful islands can offer, something rather unique and so hard to find in the noisy fast-moving world you have left behind.

From any of the hilltops you can see the attractive and alluring white beaches of Tean so let our boatman take us there to begin the next chapter.

Kestrel

Gannet

Puffin

41

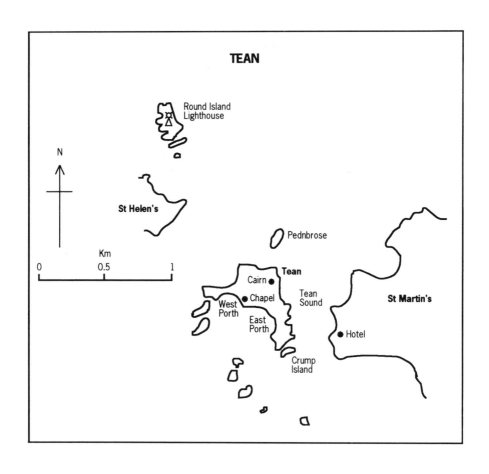

TEAN

Round Island
Lighthouse

N

St Helen's

Km

0   0.5   1

Pednbrose

Tean

Cairn ●

Tean
Sound

St Martin's

● Chapel

West
Porth

East
Porth

● Hotel

Crump
Island

Oystercatcher

4

TEAN

Old Walls, Tean

I remember the first time I saw Tean. It was just after dawn on a brilliant summer day in June and I had walked up from Higher Town on St Martin's and over The Plains. Suddenly the grandest sight in all of Scilly met me as I topped Tinkler's Hill. I just stood there and stared, mesmerised at the beauty of it all. To my left all the Eastern Islands set in a silvery sea sparkling in the early morning sun. Ahead of me lay the island of Tean standing out with all its magnificent colours of green and brown.

The tide was low and Crump Island on the south side of Tean was surrounded with sand, brilliant white at the top of the beach where it was dry, turning to grey further down. A flotsam of seaweed lay around the rocks and shoreline, contributing the colour of burnt umber to the scene. An oystercatcher piped a 'pic-pic-pic' amongst the lichen-covered rocks on the foreshore above Tean Sound, that short stretch of water that gives shelter under the lee of the island but has tremendous surge currents at certain states of the tide. Little boats, red, blue and white, lay at anchor in the sound and seabirds of many species flapped lazy wings as they made their way to fishing grounds nearby.

On Tean itself dark green humps of ivy-covered walls and buildings stood out suggesting dwelling places of long ago. Rocks, cliffs, sand and shore with acres of green slopes were all there to be explored and investigated. Beyond Tean lay ever more beguiling islands, St Helen's, Bryher and Tresco. To the north lay Round Island, a huge bastion of rock on top of which was built a lighthouse perpetually winking every ten seconds. Around it, radio masts and outbuildings of concrete indicated the days not so long ago when the station was permanently manned by Trinity House keepers.

Tean has always been associated with the Nance family, who first inhabited it in 1684 and started up the kelp industry. The remains of their cottage can still be seen between East and West Porth beside the ancient Christian chapel of St Theona and its 16 stone-lined cist graves. Nearby is a Romano-British midden or waste tip of the sixth or seventh century. An old well with a nineteenth-century rusty iron pump and two huge iron water containers can still be found deep amongst the tall bracken in the centre of the island. This was used as a source of water for cattle, which grazed on the rich vegetation until 1945. Some families from St Martin's used to live on the island in summer, looking after the animals and collecting the rich harvest of seaweed for kelp-making; their dwelling places can still be seen surrounded by tall pink-purple *foxgloves*. The cattle were herded across Tean Sound at low tides, when they were made to swim the short distance to the island.

Just to complete this brief history, a Bronze Age grave can be seen on the summit of the highest hill to the north-east of the island, and at very low tides ancient boulder walls and a large cairn are uncovered in West Porth.

Many species of seabirds breed on Tean. My favourite must be the *oystercatcher*, which can be seen at any time of the year, its whistling cries giving a haunting and wild quality to the shoreline and rocky beaches. A bird of medium size, some 40 centimetres long, the oystercatcher is conspicuous and striking, with its long orange-red bill, stout pink legs, and black head and back contrasting with its pure white underparts. On Tean oystercatchers can often be seen in great numbers on the beach at East Porth. Sit quietly on the trefoil-covered bank when the tide is low and watch large flocks alighting, making quite a clamour with their shrill piping calls. Watch them advance and retreat with the small lapping waves and push their beaks deeply into the wet sand, looking for worms and shellfish. Despite their name they do not catch oysters but they are particularly adept at breaking into mussels. Their other common name, 'sea-pie', is a more suitable description of them. As you contemplate this peaceful scene, be sure you do not move, for at the slightest alarm the whole flock will rise and fly away with a strident 'kleep, kleep, kleep' repeated by many individual voices.

The oystercatcher makes its nest in a variety of places: amongst shingle or sand, often where a little dead weed and drift has collected; in vegetation patches on rocky headlands and islets, or on almost bare rock. The nest itself is just a bare scrape, often unlined, or with a few pieces of thrift flowers, shells, pebbles or rabbit droppings. The two to four cream-coloured eggs are blotched and spotted with dark brown or grey and are laid from April to June. The parent birds take turns to incubate them and are particularly anxious when the chicks hatch out after 27 days. Any intruder will be met by the adults flying around and whistling in alarm, while the chicks crouch and vanish among the stones and vegetation. Often, when disturbed the sitting bird will try to draw you away from the nest site by shuffling across the ground, feigning a broken wing. If you persist in hanging around, the bird may well fly fast and low from behind and give you a sharp blow on the head with its bill or feet. A very unnerving experience, which will be repeated with increased boldness and ferocity until you withdraw. It is amazing how formidable such a delightful bird can become when it is defending its young.

I remember one interesting occasion when I saw two *carrion crows* flying overhead. Suddenly an oystercatcher rose from the nearby heathland and chased after the crows; I watched as one of the crows broke away and dived straight down onto the ground where the oystercatcher had been. The other was by now some distance away, hotly pursued by the angry bird. Suddenly I realised the significance of what I was witnessing and raced across the ground, but too late; the crow flew away, leaving the remains of two oystercatcher's eggs dripping out their contents onto the scrape in the earth. Crows can be cunning predators and on this occasion they were clever in hunting as a pair. Luckily, as with others of their species, the carrion crow is uncommon and does not often breed in Scilly.

Scilly is in a strategically placed position to be used as a resting place for many birds during the spring and autumn migrations, and although over 400 species have been recorded on the islands, only about 50 regularly breed. One of the commonest birds breeding on just about all the islands is the *wren*.

About 10 centimetres long, the wren is one of the smallest of British birds, only the goldcrest being smaller. You will encounter this familiar bird, with its plump brown body and short upturned tail, just about anywhere on Tean, amongst the brambles and bracken and on the rocks and stone walls near the beaches. The wren on Scilly is slightly different from that on the mainland, being rustier in colour and having a more pronounced 'eyebrow'. It is also less secretive in its habits and more easily seen. The wren has an amazingly loud, joyous song, a prolonged, breathless jingle of strident musical notes and high trills. It seems incredible that such a volume of sound can emanate from such a small throat. The cock wren will build several nests up to 3 metres off the ground in almost any nook and cranny, but especially in ivy growing on walls. The nest is domed, with an entrance only 2 centimetres in diameter, and is made of moss, grass and leaves. The hen bird chooses the nest in which to lay her eggs and lines it with feathers. The eggs are laid from early April to July and several broods are reared each year. Males sometimes have several mates, each of which lays four to eight small white eggs sparingly spotted with reddish-brown. These are incubated by the female and hatch out after about 14 days. The young are fed by both parents and fly out of the nest in about 15 days. On Tean, I have found wrens' nests among the ancient boundary walls and in the ivy covering one of the old dwellings. Wordsworth's poem, 'A Wren's Nest' is very apt even though there is no brook on Tean

> These find, 'mid ivied abbey-walls
> A canopy in some still nook;
> Others are pent-housed by a brae
> That overhangs a brook.

Wrens are very sensitive to cold weather and in winter they often huddle together at night to keep warm. The comparatively warm winters in Scilly probably help to explain why the bird is the commonest breeding species there.

Round Island, just a kilometre to the north-west of Tean, is the home of many seabirds. One of the more interesting of these is the *storm petrel*, formerly known to sailors as Mother Carey's chicken because it follows ships flitting behind them over the rough water of their wake looking for food. This small bird, only 15 centimetres long, has sooty-black plumage with a conspicuous white patch beneath its tail. Swallow-like, it skims over the sea, appearing almost to walk on the water, pattering briefly on the surface with dangling black feet. It spends most of the year far out at sea, and is truly an oceanic bird, only coming back in April to remote islands to breed. The storm petrel, like others of its species, is a buoyant swimmer and gathers in huge flocks at rich fishing banks to feed on the larger planktonic animals and smaller fish.

The storm petrel breeds in large colonies, the individual nests being just a few blades of dry grass placed in an old rabbit burrow or in holes in cliffs, or, as on Round Island, under large boulders. A single white egg, with small dust-like reddish-brown spots at the larger end, is laid from June to August. The young chick hatches out in 38 days and leaves the nest after a further eight or nine weeks. The parent birds are very seldom seen as they fly into their breeding places at dusk. Incubation is undertaken by both sexes and the changeover period, usually at dusk, is carried out every two or three days amidst the sound of purring noises coming from the nest sites. Imagine the scene on a warm June night, the stars in the sky just beginning to show as the last orange glow of the sun fades over the distant horizon. You wait patiently for the purring noises to begin, with only the winking of the lighthouse to distract your concentration. The rising and falling noises begin, and suddenly, without warning, the air is filled with flittering bat-like forms. If you could see them the incoming birds would be plump and glossy from exercise and feeding at sea, while the outgoing bird would be thin, with an empty stomach, anxious to escape to the ocean to feed and clean off the earth and accumulated dirt from its body. The storm petrel is a very common breeding bird in Scilly, particularly on Annet and the northern group of islands, and although quite agile when on land in its breeding environment, it is the air and the ocean waves which it loves and where it is really at home.

In your wandering over Tean you will notice how fertile much of the land is; of course, for many hundreds of years the island was farmed, but the areas formerly cultivated are now overgrown with a tangle of bramble and bracken. You will not fail to notice the abundance of *hogweed*, which grows where the soil is deep. This tall, robust plant has a thick hairy stem, on top of which sprout the umbrella-like arrangements of stalks, each of which bears a mass of small white flowers. These bunches of flowers are flat, which distinguish the plant from many other similar species. The hogweed sometimes flowers in winter but is best seen in late summer, when it is the commonest of its family (the Umbellifers) around. The dead, dry hollow stems stand stiffly upright all winter and make wonderful hiding places for hibernating beetles and snails. Young hogweed shoots are very succulent when cooked and eaten like broccoli.

Another common plant to grow on Tean is *red clover* and this is probably a leftover from the time when it was grown here by farmers as an important ingredient in the hay crop. Its flower heads at first are round, afterwards becoming longer than broad; they are purplish-red in colour and bloom from May to September. The large elliptical leaflets are frequently marked with a whitish band the shape of a quarter moon.

The flora of the island's beaches and dune grasslands is truly wonderful and some of Britain's rarest plants are to be found here. The *dwarf pansy* for instance does not grow on the mainland and is found only on a few of the beaches in Scilly, where it frequents sand dunes, particularly where the ground has been disturbed by rabbits. The plant is small in all its parts, being only 4 to 8 millimetres tall, with creamy-white flowers, and it is not easy to pick out amongst the other small flowers of the short turf. Look for it at the end of March to early May on the sandy turf about West and East Porth.

Two rare plants of the Spurge family are to be found on the beaches of Tean: namely, the *sea spurge* and the *Portland spurge*.

Look for the sea spurge on the higher parts of sandy beaches where they give way to the dunes. Up to 25 centimetres tall, this plant has narrow, concave, very thick leaves arranged in whorls around a single stem, giving a columnar appearance. The yellowish-green flower heads appear from June to September. The Portland spurge is similar to the sea spurge but is shorter, only some 15 centimetres tall, has branched stems and has often a reddish or purplish hue to the whole plant, especially in its lower leaves. Although growing in similar places to the sea spurge, it also favours the more exposed places on cliffs and rocky slopes. Like all spurge plants, these two species exude a white poisonous sap when the stems are broken or cut.

Another plant of the sandy beaches is the *sea holly*, quite unmistakable with its large prickly waxy-looking leaves, and compact thistle-like heads of misty blue flowers. The plant grows to some 60 centimetres in height and the flowers appear in July and August. Sea holly has become rarer over the years, possibly because it has been picked and even uprooted for use in flower arrangements. It has a large system of fleshy roots, which were once candied with sugar and used as sweets, called 'eringoes', after its Latin name. These were eaten as a medicinal rejuvenator and stimulant. Colchester, on the east coast mainland, was famed for this product in the seventeenth century.

Whenever one comes across sea holly, look also for the attractive *sea rocket* with its pale lilac-coloured flowers and somewhat bushy appearance. Up to 45 centimetres in height, it has fleshy lobed leaves and its flowers, which bloom from June to September, are often covered with insects attracted by its sweetly smelling scent, especially in autumn, when it may be the only remaining plant flowering on the shore. It is an erratic plant and sometimes it is abundant on certain beaches one year and then disappears altogether for a period, but I have always managed to find it on the sandy shores of West and East Porth, as well as on Tresco and Samson.

One cannot fail to mention two of the outstanding botanical features of Scilly, the *rock spurrey* and the *rock samphire*. Both these plants flourish in conditions of high concentrations of salt and are equally at home on isolated sea-swept rocks or beside walls on the larger islands. They are abundant on all the islands, so you will have no difficulty in finding them.

The rock spurrey has a thick woody rootstock from which sprouts a large number of long trailing stems clothed with glandular hairs. It has thick bright yellowy-green leaves, but its most attractive feature is the abundance of neat little flowers with deep pink petals which bloom from May until September.

The rock samphire cannot be described as a beautiful plant but seen here on Tean growing all over the rocks above high tide it has a quality all of its own. Its colours of grey, green, yellow and pale brown seem to harmonise so well with those other hues of nature, the blue sea, the brilliant white sand and the dark brown earthy banks above the shore where they are eroded away by wind, rain and sea. Rock samphire is easily recognisable as the only Umbellifer plant with thick, succulent leaves which are divided into many narrow, spoon-shaped segments which taper at both ends. The flowers, which appear from June to September, are yellowish-green and grow in small stiff clusters which unite to form a large flat-topped umbel. The whole plant, about 40 centimetres high, is exceedingly smooth and is covered with a greyish-green 'bloom' that is characteristic of so many seaside plants. Samphire has been used as a vegetable, chiefly in pickles, and in the eighteenth and early nineteenth centuries it was important to the economy of some Scillonians. It was collected and put into small casks and covered with a strong brine. It was then exported to many places, including London, where wholesalers would pay up to 4 shillings a bushel for it. The samphire of Scilly was much sought after and was praised for its size, quality and abundance and was thought to be far superior to that in Cornwall.

There are several species of the Dock family to be found on Tean, two of which are the *curled dock* and *fiddle dock*, each having their own botanical differences, mainly to be found in the nature of their fruits. On the sand dunes and rocky shores look out for the curled dock, which is the tallest of our common docks and is remarkable for the way the dying stems of one year's growth for a protective tent over the new shoots from the sea spray thrown up in winter gales. The fiddle dock has branches which grow out almost horizontally from the main stem and is more often found in the waste places well above the beach.

One of the world's rare docks, the *shore dock*, only found in Western Europe, has its main British outpost in Scilly. Here it is found on most of the islands, growing on the sandy and rocky shores, but as in south-west England and South Wales, it is in danger of being replaced by the more prolific and ubiquitous curled dock. You may find it on Tean, so look out for the distinguishing feature of its oblong grey-green and blunt leaves.

Docks have always been used as an effective antidote for nettle stings, the leaves being rubbed hard against the skin so that the juices are squeezed out onto the swollen tissue. Since ancient times the leaves of the dock have been used to keep butter cool in the summer, and indeed the *broad-leaved dock* was known as 'butterdock'. Interestingly, this common species, always associated with human activities, is common on all the inhabited islands but is absent from all those that are uninhabited, like Tean.

Tean is an interesting place for butterflies. Some species of butterfly that do not regularly migrate are weak fliers and remain localised in one area or on one island. They seem unable or unwilling to fly across the water and form new colonies; so on remote islands they develop interesting subspecies through inbreeding. On Tean the common blue butterfly has evolved into a separate subspecies known as the *Tean blue*. Here the female has an extensive scattering of silvery-blue scales on its upper hind wing. In addition, both sexes have a characteristic variation in the spotting on the underside of their hind wings. This distinct variety of blue was discovered in 1938 by the late Mr E.B. Ford and Mr W.H. Dowdeswell, who found that this Tean blue had not crossed over the short distance of about 150 metres to St Martin's, where there were large numbers of the common blue. So here we have on the lovely island of Tean a butterfly subspecies probably found nowhere else in the world. The reader may spend many a happy hour amongst the yellow *bird's-foot trefoil* plants growing in profusion above the beach at East Porth, photographing this attractive butterfly, but you will have to have patience for the butterfly is very active and is easily disturbed.

Mr Ford also discovered that Britain's commonest butterfly, the meadow brown, is also represented in Scilly by a distinct subspecies, the *Scilly meadow brown*. This butterfly has distinctive markings in the form of an extra pale band on the underside of the hindwings. It is interesting to note that this subspecies is characteristic of all the meadow browns on all of the islands; this is not altogether surprising as the meadow brown is an active and fairly strong flier and has obviously managed to enlarge its range all over Scilly. Indeed, it is so active that, unusually for butterflies, it can be seen flying on dull days and even in drizzle. You will have no difficulty in recognising this butterfly as it flits amongst the grasses and rough areas on Tean. The female is larger than the male and is more brightly coloured as well. This is in contrast to most butterfly species, where the male is the more colourful. The meadow brown has a single prominent false eye (occasionally two) on each forewing on both the upperside and underside. These eye-spots stand out well against the orange and brown wing colours and are said to confuse predators. Birds, for instance, will tend to peck at these spots rather than the real eyes, thus allowing the butterfly to escape with just a damaged wing. All the family of brown butterfly species, including the *speckled wood, grayling* and *small heath*, have such false eye-spots, each with a greater or lesser degree of prominence. The female lays her eggs singly on grass stems in summer, and these hatch out into green and yellow caterpillars after about two or three weeks. The caterpillars feed on grasses at night during mild spells throughout the winter. They form pale green chrysalids with black stripes at the base of grass stems in May and June, from which the adult butterflies emerge after about a month. Like most butterflies, the meadow brown has a short lifespan of only three or four weeks. As you wander over the islands, be it Tean or just about anywhere, the Scilly meadow brown will rise up out of the grass in large numbers and fly around in front of you.

When the first autumnal rains begin in October, the short rabbit-cropped grasses above the beaches become damp and there is just that hint of decay about, then fungal mycelia begin their work and the first fungi begin to appear. Look for the *common puffball* fungus, which is snow-white at first, gradually becoming brownish as it ages. From 3 to 6 centimetres across, it has almost no stem at all and has to be carefully teased out of the grass. This is an excellent fungus to fry up and eat with onions, especially when its flesh is still white throughout.

Autumn, too, is the time when in the early morning the undergrowth of bracken and brambles becomes festooned with dew-laden webs of spiders. Many of these belong to one of the commonest of British spiders, the *garden-spider*, known also as the garden-cross and diadem spider because of the white cross on its back. It spins a complex web which sometimes spans several metres. It is amazing how the female spider manages to bridge a huge gap between two distant supports with a single strong strand that can withstand wind and rain. It does not do so by laboriously crawling along the ground and climbing up the support. Paying out a single thread without entangling it would have been an impossible feat. What she does is to let out a gossamer thread, which drifts with the wind until it snags on a distant object; then the spider walks along it, coiling it in and eating it, while a new much thicker thread is paid out from her body. After this strong support thread is in place then the spider will spin the remainder of the web until finally it can rest in the small inner spiral of the complete structure, waiting for her prey. When an insect becomes caught in the web, the spider feels the vibrations through its feet, rushes to the victim, bites and paralyses it and then wraps it in a silk cocoon. Finally, she injects it with enzymes to turn its body into liquid for eating. She avoids sticking to her own web by coating her feet with an oily substance. When the web becomes too damaged to work efficiently, she will eat it and use the material again. The male spider, only about a quarter the female's size, does not share her web or help in its making, but sits casually nearby and scavenges off its victims. In the autumn up to 800 eggs are laid in a single mass, protected by a layer of yellow silk, on the side of a rock or on a stem; the young spiders hatch out in spring and at the slightest disturbance the mass of tiny spiders disintegrates and they all scatter.

As you leave Tean and its history and wildlife you may see one of the most magnificent and lovely sights of the islands: an autumn sunset where, as that great star gradually lowers over the grandeur of rocks and sea, the colours of the sky change from shades of blue and pale yellow to deep orange and red. The rocks stand out as solid black bastions and passing gulls fly across the flame-red disc, calling their notes of raucous laughter. Almost imperceptibly the sun disappears and the long shadows of the islands merge with the enveloping darkness.

Autumn changes to winter and Tean is left unvisited, quiet, alone and with a sombreness that is not too melancholy, for although many of its birds have migrated to distant places, there is still a great sense of well-being and of permanence. You know that the whole island will awaken in the early months of next year when the sun starts to climb remorselessly higher and higher as each day passes, and the warm Gulf Stream keeps away the cold frosts so common on the mainland. The hibernating insects will soon be awake and busy about their business and the birds in their thousands will be returning from their winter quarters.

St Helen's, our next island, is nearby and overlooks Tean just to the north-west.

Round Island

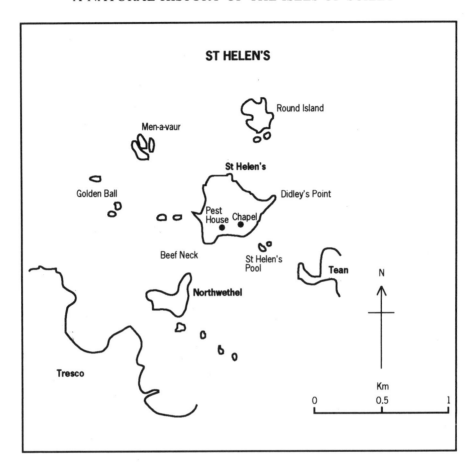

Guillemot

5

ST HELEN'S

The Pest House

There is only one landing point on St Helen's that offers a relatively easy disembarkation onto the shore. An old stone quay still exists on the south side of the island between the Pest House and the sandy beach, its debris of boulders and dressed stones, which appear as the tide lowers, gives the only clue that this was once the main landing place for the island. When it was first built and for what purpose I could not discover but it was probably constructed in about 1764 to assist in the building of the Pest House in that year. The Pest House was put up as an isolation hospital to accommodate those people from visiting ships who were suspected of having diseases such as cholera, typhus and yellow fever but it was never actually used. The hospital now survives as a single-storey granite wall to roof height with a brick chimney, and although the roof has long since gone, the main room and two small rooms are still intact; none of the rooms interconnect but there is a blocked window between the smaller ones. Timber lintels, a stone fireplace and traces of wall plaster give a certain atmosphere and character to this desolate and somewhat eerie place.

Just to the east of the Pest House can be found the ruins of the hermitage and chapel of St Elidius. Legend has it that Elidius, a holy man who lived here in the tenth century, came across one

Olaf, a warrior 'king' from Norway, lying badly injured and dying after being shipwrecked on the island. The year was AD 993 and gradually Elidius nursed Olaf back to health, at the same time converting him to Christianity; Olaf sailed back to his homeland and spread Christianity to the whole of Scandinavia. Elidius was then canonised and in the eleventh century a monastery was added to the site, which became a destination for pilgrims throughout Europe. Sadly this monastery fell into disuse in the fifteenth century and was in disrepair by the Reformation. The ruins of this complex ecclesiastical site can still be seen amidst the brambles and bracken that threaten to engulf and obliterate all that remains of it. A plaque giving details of the hermitage and its history has recently been erected.

Sit amongst the chapel ruins on a bleak, windy and wet day and look across the remnants of an ancient field system with boulder walls and stony banks to the sea, now dark and foreboding and whipped up into white spray by the westerly gale. Just imagine what it must have been like for the lonely Elidius, living without comforts in the circular stone-walled cell he called his home. What thoughts would have passed through his mind? What would he have eaten and drunk, and who would have been his visitors and friends? His faith must have been exceedingly strong to sustain him through the travails of life in such a place. You look across the tangle of foliage, wet and dripping with the soft rain of Scilly, to the five graves, each marked with a granite stone, and suddenly realise that the saint himself lies but a few feet away.

In early summer you will not find these ruins at all peaceful for you will be harassed by the many *lesser black-backed gulls* that have taken residence on and around them. Only a few years ago in the early 1990s there were a large number of brown rats on the island and because they presented a threat to the small puffin colony on the island, poison was put down, and by 1995 they were largely eliminated. From my own observations, when the rats roamed free, the black-headed gull population was kept in check, but after the rats were eradicated, the gull population exploded. During my walk around the island in June 1997 it was almost impossible to avoid groups of these gulls in various stages of their breeding cycle, some with eggs and some with young. They were mainly concentrated around the Pest House and the hermitage site, although there were also large colonies on the western side of the island. Did the rats keep the gulls in check or was it just coincidence that after they were gone, the lesser black-backed gulls increased in number so markedly? Interestingly on Tean, where the rats still roam unpoisoned and unmolested, the black-backed gull numbers have remained stable and quite small.

The lesser black-backed gull, of which there are about 3,800 pairs in the islands altogether, is about 58 centimetres long, and with its yellow legs and feet is easily distinguished from the herring gull and greater black-backed gull, both of which have flesh-coloured legs and feet. The nests of the lesser black-backed gull colonies are easy to find scattered amongst the thrift-covered boulders and bracken-shrouded slopes. A small scrape in the ground is lined with seaweed and grass on which the two to four olive-brown eggs, spotted and blotched with grey and darker brown, are laid from April to June. Often there is no nest material and the eggs are just laid in a hollow on the grass or bracken. The chicks hatch out after about 26 days and are born with their eyes open; they are able to walk almost immediately, and when danger threatens, signalled by an angry 'kyeok' alarm cry of the adults, they try to hide in a nearby crevice or crouch under a tuft of vegetation. Meanwhile, the adult birds will continue to harry any intruder by diving down at it and hitting it viciously with its legs. The small chick is especially vulnerable to predation by other gulls, including its own species, and should it wander far from its nest it is unlikely to survive, and even when it returns home to its nest, it may be killed and eaten by its own parents. This cannibalism is common amongst the lesser and greater black-backed gulls and herring gulls but is less so in the case of other gull species. Unlike the other two species, the lesser black-back is a migrant and in autumn it flies south, reaching Spain and the north and west coasts of Africa by December. It is not seen in Britain again until March the following year.

Slightly smaller than the previous species, the *herring gull* also nests on St Helen's, as indeed it does on most of the other islands. Its call of 'hau-hau-hau' echoing around the rocks and little coves is a familiar one, and is very similar to that of the lesser black-backed gull. The herring gull nest is usually more substantial than that of the lesser black-back and is made up of seaweed and pieces of turf lined with grass, but occasionally it is only a barely lined scrape. In Scilly the herring gull nests singly or in small groups, the nests more often being found close to the high-tide mark, on the cliff tops or on rocky outcrops near the shore. The two or three eggs are laid from April to June and are indistinguishable from those of the lesser black-backed gull. The young hatch out in 26 days or so and their behaviour is very similar to the previous species.

In the winter many of the breeding gulls of Scilly, such as the fulmar, kittiwake and lesser black-back, will be absent but others, such as the *black-headed* and *common gulls*, migrating from the north, will take their place. Enjoy them all, the gulls and their laughing calls and rejoice that they should give so much pleasure to so many with their antics, wheeling and darting over the black weed-coated rocks, white sands, and blue sea and sky. We would miss them if they were not here.

To the north-west of St Helen's there is a rocky, boulder-strewn bay half encircled by a 10-metre-high cliff of hard brown earth. Here the soil and clumps of thrift, grass and *Hottentot fig* tumble down onto the beach as the fertile slope of the island is gradually eroded away. In this cliff face a colony of some 20 pairs of *puffins* have established themselves by burrowing into the soil, each building a nest which is just a scrape at the end of a half-metre to 2-metre-long tunnel. Occasionally this scrape is lined with grass and feathers.

The puffin is one of our more fascinating and charming seabirds; about 30 centimetres long, it has a white breast and cheeks, with black wings, black back and head crown. Its main distinguishing feature is its triangular red, blue and yellow laterally-flattened bill and bright orange feet and legs. A member of the Auk family, the puffin has a clumsy appearance, almost clown-like, as it stands on its webbed feet looking out to sea amongst the bright flowers of the coast. Like others of its species, the puffin is migratory, spending most of its time at sea, only coming to land in March to breed and leaving again for the open ocean in August.

The puffin is a gregarious, social bird and during the breeding season, especially in the evenings, the adults sit about and parade on the cliff tops in a sort of recreational assembly. The single white egg in the protected and well-sheltered nest seems to come to no harm by being abandoned for a few hours. Now and again one or two birds will take wing, flying off with rapidly beating wings and a flash of orange feet, to join razorbills floating on the waves below the towering rocks of Men-a-vaur. The puffin is usually a silent bird but occasionally it will utter a 'arr-arr-arr' during the breeding season. Both parents take turns incubating the egg; the young hatches after about 41 days and is fed by both parents with fish. One adult can carry as many as 30 small fish in its enormous parrot-like bill and, contrary to popular opinion, the fish are not carried head to tail alternately but are carried haphazardly across the well-serrated beak.

Six weeks after laying their eggs the adult puffins leave land altogether and return to the open sea. The deserted chicks are left to fast in their burrows until after about a week, when their wing and tail feathers have formed and realising their parents have gone, they flutter down onto the beach and into the sea. This act normally occurs at night or before dawn, when predators like the black-backed gulls are inactive. Once in the sea the young chicks paddle furiously (for they cannot yet fly) out into the open ocean and, if attacked, they dive down below the surface. It is interesting to note that as far as is known the parents have little or no contact with their offspring, certainly at this stage, and it would appear that, given the darkness inside the burrow and their silent nature, they have little chance of recognising their own anyway. Young puffins, like so many other migratory birds, have to find their way alone to their wintering ground. Little is known of where they go or how they get there; it is thought that they spend their time in the middle of the North Atlantic feeding on plankton and small fish.

Puffins provided a valuable source of meat in the Middle Ages and in 1337, when the Isles of Scilly were included in the Duchy of Cornwall, 300 puffins were paid to The Black Prince (son of King Edward III) as part of the rent by the tenant-owner, Ranulph de Blanchminster. At that time they could be eaten in Lent, being classed as fish!

The towering rocks of Men-a-vaur sticking out of the sea looking like a great citadel provide huge cliffs which are an important seabird breeding site, especially for the *razorbill* and *common guillemot*. Like the puffin, these members of the Auk family are essentially maritime birds, spending most of their time in the deep ocean, coming only to land in the breeding season, and then being restricted to cliffs and rocky ledges.

The razorbill is about 40 centimetres in length, with black upper parts and white breast, and its laterally compressed bill is crossed midway by a prominent white line. The only British bird with which it could be confused is the common guillemot, but the latter has a slender pointed bill and thinner neck and its back is usually dark brown in colour rather than black. Both species shun the shoreline and beaches and are to be seen around Men-a-vaur and other outlying islands, swimming amongst the waves and riding up and down with the ocean swell. Both species are very sociable in the breeding season, gathering in great numbers on the rocks and cliffs. The guillemot favours the bare cliff ledges on which to lay its single egg, but the razorbill will make its home amongst boulders and tumbled stones, where its egg is laid in a crevice or hollow beneath a rock and is not usually to be found in the open or on a ledge. Neither bird makes a nest and their eggs are usually laid in May and June. The colour of both the razorbill's and guillemot's eggs vary enormously; those of both are heavily blotched and streaked with shades of black, brown and red but those of the guillemot have a bluish-green base colour rather than the dull white of the razorbill. The shape of the guillemot's egg is interesting for it is markedly pear-shaped and when touched, it tends to turn on its axis rather than roll. This prevents it from falling off the flat cliff ledges onto the rocks and sea below. Because guillemots nest in such large numbers in close proximity to one another on unprotected cliff ledges, nature has clearly evolved in such a way to protect the species. Even the wide variety of their egg colouration possibly helps the parent birds to recognise their own, although after only a short while they usually become very dirty on the insanitary and crowded ledges.

The fledging period for young guillemot and razorbill chicks is less than a third of that of the puffin, and clearly because of their open position they are more susceptible to predation by other seabirds. Therefore the most successful birds will be those that are hatched in the shortest time and which spend the briefest period as chicks on the cliffs. Again, evolutionary processes over a long period of time have helped these species to survive in a hostile environment. The young razorbill and guillemot chicks grow rapidly and soon have a soft loose coat of feathers covering a layer of thick down. They grow restless on the ledges and after about 15 days, and usually at night or early in the morning, they descend on a long gliding flutter to the sea. Occasionally if the tide is low they can't quite reach the sea and bounce off the rocks, but they are seldom seriously hurt and manage somehow to scramble to the safety of the sea. Sometimes, too, the weather is bad, with heavy seas breaking, and the birds have difficulty in making headway to the open sea, being buffeted and thrown back by heavy surf against the rocks. But their agility in the water is surprising and with encouraging calls from their adults farther offshore, they paddle furiously towards them, anxiously squeaking in response. Once united, the adults lead the chicks out into the open sea; how different from that of the puffin!

So we leave the world of the auks, those salt-water diving birds with their large webbed feet, their distinctive stance and whirring wing flight, who give us so much pleasure just watching them in their magnificent surroundings of rugged cliffs and ocean waves. We shall meet them again on other islands.

St Helen's has a marvellous flora, not so much in variety, but in display and colour. *Thrift* or *sea pink*, that most beautiful and well known of seaside plants, covers the rocks in mounds of pink, and *common bird's-foot trefoil* gives a welcome to the common blue butterfly.

Thrift is the most familiar plant of the spray-washed banks and cliffs above the beaches. Its rootstock is long, much branched, thick and woody, and every branch ends in a bundle of fleshy, narrow glaucous leaves. From each bundle rise a number of leafless and softly hairy flower stems some 15 to 20 centimetres long. Each stem bears a cluster of closely packed, rosy-pink flowers which bloom from April to October; but its most glorious display is seen in May. Long after the beautiful flowers have withered and died, their outer remains survive as papery sheaves. These clusters of pale brown give a certain quality to the bleak landscape in midwinter, contrasting well with the blue-green cushions of the living plant itself.

Common bird's-foot trefoil covers many parts of the island in a blaze of yellow from June through to September. From a short woody rootstock originate several tough trailing branches which are angled and twisted. The leaves are divided into five oval leaflets, the bottom part often being bent back, leaving the form of a 'trefoil'. The name 'bird's-foot' comes from the appearance of the seed pods which make up the shape of a bird's claws. The flower heads, containing two to eight flowers, are borne at the end of 10-centimetre-long wiry stems. The flowers themselves are a pretty yellow tinted with red, and on St Helen's the red colour in some places is particularly marked, giving the whole plant a very striking appearance.

On the short turf on the north-west of the island grows large matted tufts of *English stonecrop*. It makes an attractive decoration on the bare rocks and amongst the cushions of thrift, especially in May and June when its pink or white star-like flowers are in profusion. The thick fleshy leaves of this plant are bluish-green or red, or a mixture of the two. You may also come across the much rarer (in Scilly) *biting stonecrop*, easily identified by its golden yellow flowers, which appear in June and July. It is more commonly found in sandy places near the shore on the south of the island.

Before the Second World War much of the higher ground on the island was covered with *ling* and *bell heather*, but much of this was destroyed by fires caused by incendiary bombs dropped by enemy aircraft in 1940. Further fires raged across the island in 1949 during a period of extremely dry weather. Both heathers can still be found here, bell heather being the first to bloom at the end of June, with its reddish-purple flowers shaped like small bells and clustered around the ends of woody stems. Ling, as described in Chapter 2, has sprays of smaller, less reddish, and more lilac-coloured flowers.

The most characteristic plant of St Helen's for me must be the *Hottentot fig*, or *mesembryanthemum*. It is unmistakable, with its large yellow or purple flowers, thick woody trailing stems and long fleshy leaves. Introduced by Augustus Smith on Tresco in the middle of the nineteenth century, this native plant of South Africa has now established itself on most of the islands. Its stems are widely used by gulls as nesting material and this is the main reason for its rapid spread, and it now smothers out many native species on the cliff tops. The main enemy of the Hottentot fig is frost and after the rare occasions that this occurs in Scilly, long ropes of dead stems and leaves are to be seen on the exposed places where the plant normally thrives. There are usually always parts of the plant left alive, however, and it is not long before the plant recovers and the effect of frost damage disappears. The flowers are rich in nectar and are much visited by insects, and to me one of the great attractions of St Helen's is to sit on the summit amongst this flower and listen to the gentle hum of bees as they gather their food from flower to flower. You can eat the fruits of this succulent plant and they have a pleasant flavour when ripe.

All around the island on the rocks and cliffs by the shore you will find the *common scurvy-grass*; this plant has a thick woody rootstock from which spring many trailing and ascending hairless, angular stems. The shining bright green fleshy leaves are heart-shaped and form a loose rosette at the base of the plant. They are full of vitamin C and it was common for sailors in the days of sailing ships to gather up the plant to eat as a protection against scurvy. Captain Cook made reference to this practice in his diaries. The small white four-petalled flowers bloom from March until July and are one of the first coastal plants to come into bloom. It is usually only some 10 centimetres high but where it is found in damp and shady places exposed to salt spray it can be over 30 centimetres tall with exceptionally large leaves.

Another plant to look out for is the *sea storksbill*, which grows on trampled paths, on short grass and on bare ground where the soil is shallow. It seems to grow well where there are high levels of nutrients from the droppings of gulls and rabbits. It is a low-growing, attractive little plant which can easily be distinguished from other storksbill species by its small, oval leaves with rounded lobes around their edges. The whole plant is covered with soft hairs and feels slightly sticky. It flowers from May to September but the five pink or white rounded petals of the flower drop off early and are often absent altogether. The sea storksbill is abundant on nearly all the islands, as is an allied species, the *common storksbill*.

The common storksbill is again hairy and sometimes sticky, but has pink or purple flowers and its leaves, unlike that of the sea storksbill, consist of pairs of deeply cut leaflets arranged on opposite sides of a common stalk. This plant grows commonly amongst the sand-blown wastes on the south side of the island. Both of these species of storksbill have long pointed fruit each containing five seeds; the shape of the fruit, like a stork's long bill, gives the species its name. Allied to them are the *cranesbills*, some of which are also common in Scilly.

As you walk around the old ruins of the chapel, look for the 60-100 centimetres tall, square-stemmed plant of the *balm-leaved figwort*. It is a downy plant with heart-shaped, wrinkled and toothed leaves. The globular red-brown flowers at the end of branched clusters are only a couple of centimetres across and bloom from July to August. A rare plant on the mainland, the balm-leaved figwort is abundant on most of the islands. There are several other species of figwort but of these only the *common figwort* can be found on Scilly and then only on St Mary's, Tresco and Bryher. The roots of figwort were used by ancient herbalists as a cure for piles and glandular swellings, hence its Latin name *scrophularia* and the derivation of the English word scrofula. I wonder if its association with human habitations on Scilly has any connection with its herbal properties?

In June and July butterflies such as the *common blue* can be seen flitting around the bird's-foot trefoil, and *red admirals* can be found taking nectar from bramble flowers. You may also come across the lovely *peacock* butterfly sunning itself on a lichen-covered rock or on a blackberry bush. The peacock is so named because the brightly coloured eye-like markings on its wings are like the spots on a peacock's tail. It has a wingspan of almost 7 centimetres and when disturbed, it accentuated these eye-spots by rapidly opening and closing its wings, making a scraping noise as the wings rub together. This scares away predatory birds, who do not hang around to see whether the eyes belong to another large predator or are just a bluff. This butterfly can be considered to be one of our most striking with its rich colourings of purplish-red, blue, yellow and black. In May the female lays her eggs in batches of a hundred or more on the underside of nettle leaves, which on

St Helen's commonly grow around the ruins of the chapel. After about two weeks these hatch out into hairy black caterpillars which, after feeding on the nettles for about 35 days, turn into chrysalids which hang down from the nettle stems on silken threads. The adult butterflies emerge in July. The peacock hibernates in autumn and winter under stones or in any suitable sheltered site and emerges on a warm day in April to start their life cycle again. The adult butterflies live for almost a year from July until they mate the following May.

The *common honeysuckle* plant grows just about everywhere in Scilly, on cliff slopes, heaths and old walls, and visitors in June and July cannot fail to be impressed by their flowers, which seem to be much larger and more brightly coloured than those on the mainland. If you are lucky you may spot a *convolvulus hawkmoth* feeding on the sweet nectar of this plant with its long proboscis. This hawkmoth is a migrant from the Mediterranean in the early summer. It is one of the most powerful flying insects to come to Britain and is capable of flying over 1,000 kilometres in just a few days. With a wingspan of 12 centimetres or more, this moth is the largest of any British insect. It is grey in colour and is difficult to see as it rests on a rock, but when in flight its large abdomen, banded in pink and black, becomes conspicuous. Occasionally early migrants may breed in Britain but if you see one on Scilly it is probably just passing through on its migration route; however, look for it at dusk or in the early evening as it hovers in front of a honeysuckle, its long tongue jabbing into the flowers.

So with the approach of evening, before you leave this island with its ancient ecclesiastical history and its varied plant and bird life, climb once more up the steep slope behind the chapel to the island's summit. It is July and the sun seems reluctant to dip behind the deep blue sea of summer, and a mackerel sky, with regularly arranged small round clouds already changing colour to delicate pink, heralds an approaching depression and poor weather. You are alone on the top, perhaps awaiting sight of the boat to pick you up, or even choosing to spend the night here with a sleeping bag in your rucksack. But you are not lonely as nature is all around; seals disport themselves in the clear blue water below, rabbits abound on the flower-strewn slopes and gulls of all descriptions wheel around in the clear, crisp air. As dusk approaches take one last look at the islands spread all about you. Tresco, with its long white beaches and tall trees, stretches out just across the water; between lies the bracken-covered island of Northwethel, which, like St Helen's, was probably grazed by goats and deer in the nineteenth century. Northwethel has a luxuriant vegetation, and although consisting mainly of bracken and bramble, there are still over 50 varieties of plants to be found on its 12 acres, including most of those already mentioned for St Helen's. Imagine the scene during the Civil War some 300 years ago, when on the seventeenth of April 1651, some 1,500 Parliamentary soldiers under Admiral Blake were encamped on Northwethel in preparation for an assault on Tresco, the refuge of Royalist soldiers. Smoke from their fires and from those also encamped on Tean drifts across the water and the desultory firing of muskets echoes around the calm sound. The attack took place at night and was successful and eventually the whole of Scilly surrendered to the Parliamentary army, the last Royalist stronghold in England to do so.

Tresco is so close by that its charm and allure captivate your imagination and interest; you just can't wait to visit and explore it.

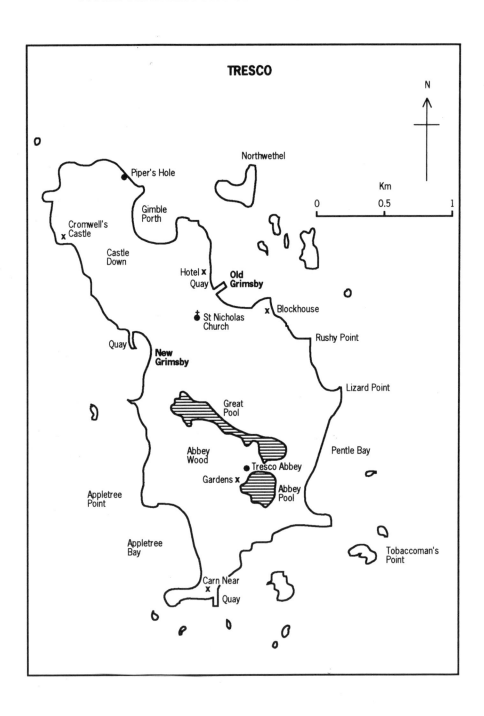

6

TRESCO

Abbey Gardens

Tresco, the second largest island in Scilly, is private and is leased under one tenancy to the Dorrien-Smith family by the Duchy of Cornwall. The underlying geology and geographical features create a very distinctive landscape with the wild heathy granite headland in the north giving way, first to the managed agricultural landscape in the central part of the island with small fields, shelterbelts and settlements, and then to the flat sandy, coastal dune systems of the south.

The influence of Augustus Smith and his successors can be seen everywhere. Flanked by Bryher to the west and by St Helen's, Tean and St Martin's to the east, Tresco is the most naturally sheltered of all the islands, but Augustus Smith knew that he would still need more shelter if his passion for delicate subtropical plants was to be satisfied. So he planted belts of cypress and tall Monterey pine from California to reduce the wind speeds and to filter the salt out of the north-easterly winds. Despite the ravages of the great storm in 1990, Tresco is still the most wooded island in Scilly, with the hilltops clothed with dense plantations which extend as narrow shelterbelts down the hillsides and fringe the well-managed agricultural land. These woodlands, with their tall pines sticking up like dark sentinels against the skyline, are a dominant feature of the island.

The importance of the Abbey Gardens, with their wonderful collection of exotic plants from all over the world, and their associated areas of fresh and brackish water, Great Pool and Abbey Pool, cannot be overemphasised. As we shall see, many of the plants have spread from the gardens and have become naturalised and the pools have become a haven for birds as well as providing a habitat for rare flowers around their edges.

On Tresco the sense and aura of ancient history is all about you; from the remains of roundhouses, chambered burial graves and field systems on Castle Down, to medieval settlements around Old Grimsby and the twelfth-century Abbey. More modern are Cromwell's Castle and the remains of King Charles's Castle near New Grimsby, the Blockhouse near Old Grimsby, and the site of Oliver's Battery at Carn Near. As you land on Tresco from one of three stone quays at New Grimsby, Old Grimsby or Carn Near, the atmosphere and beauty of this wonderful island is at once apparent. I have just touched upon the history and to go further is beyond the scope of this book. In any event, you can easily read about it in the many published guide books on Scilly; my desire now is to take you through this place of paradise and help you appreciate its flora and fauna as we go. Incidentally, a good place to stay is The Island Hotel situated in a superb setting near Old Grimsby Harbour.

The long white curving beaches are a perfect habitat for the *ringed plover*. About 20 centimetres long, this plump, lively little shore bird is easily recognised by the broad black band across its white chest, its orange bill with black tip, its white collar and the black mark through its eyes. Its upper parts are dull brown, and when first sighted, it may look like a ball of spume wafted from a wave-top and being blown along the sand. But it is a ringed plover, which runs with brief pauses, tilting its head to pick up small pieces of food, then takes to the wing to fly a little way and run along the beach once more. In flight you can easily spot its conspicuous white wing-bar, which distinguishes it from its similar counterpart the *little ringed plover*. Unlike the ringed plover, which is a common resident species breeding on most of the islands, the little ringed plover is a scarce migrant which can be seen on passage from Africa either in March and April or on its return in July and August.

How often have I watched the ringed plover in late autumn, hidden from view amongst the marram grass of the dunes above

Pentle Bay. The wind coming up from the west sweeps its salt-laden air above me, but down on the sand and well sheltered by the high banks, the birds and their immature young find shelter and rich pickings from the beach laid bare by the low tide. Watch them yourself as they run to and fro, bobbing their heads down and upending the rear of their body when they see anything worth picking up. Pairs of ringed plovers form their breeding territory from early April and this is the time when they are most vulnerable to disturbance by human presence. Indeed, because of their liking for quiet beaches, they conflict with the wishes of visitors and tourists and inevitably it is the birds that are gradually reducing in numbers throughout Britain. So whenever you come across the ringed plover on the islands, be aware of this problem, especially in the breeding season, and leave them to enjoy their peaceful surroundings and to raise their young successfully.

The ringed plover's nest is just a slight hollow scraped in the sand or shingle above the high-tide mark on stretches of sandy or shingle beaches. Generally there are no nest materials, the four pear-shaped pale cream eggs, spotted with black or grey dots, being laid on the bare ground. Sometimes the nest is lined with small pebbles or small seashells. The eggs are extremely well camouflaged in their beach environment, and great care must be taken not to tread on them. The first eggs are laid in April, with repeat second or third clutches up to the end of August, and both parents take turns in incubation. After about 23 days the young chicks hatch out, and as soon as they are dry they are led away by the parents, over the beach and away from the nest site. The chicks are perfectly camouflaged and when danger threatens they vanish as if by magic, remaining quiet and still amongst the pebbles. Often, if dogs or humans get too near to its nest or young, the parent bird will feign injury by flapping a wing and shuffling across the ground. The potential predator will follow after the apparently injured bird, which, once satisfied that the danger is over, takes to the air. After about three weeks the young can fly and are then largely independent, feeding on worms, shellfish and insects of all kinds; the parents are then able to commence a second brood. Enjoy watching the ringed plover and listening to its melodious calls, 'too-li, too-li', over the magnificent setting of a Tresco beach in early summer.

Always you will see *oystercatchers* on the beaches and for many years up to 1994 I watched *common terns* feeding their young on the safety of Green Island off Tobaccoman's Point, but for one reason or another they all left their breeding ground here in about

1995. Perhaps they will return one day or perhaps not; who knows what gives the terns their wanderlust?

The Great Pool on Tresco is a place to see the *mute swan, coot* and *moorhen*. Sit beside the track that skirts the north edge of the lake on a warm spring day when an occasional breath of air causes the surface to sparkle in the reflected sunlight. Otherwise the silence is only broken by the occasional loud but short 'tewk' of the coot, or the harsh penetrating voice 'kittok' of a disturbed moorhen. The reflections of *common reed grass*, with its tall willowy stems and feathery flower spikes, are seemingly etched for ever on the surface of the black water. Then you see it coming around the corner, its wake causing small wavelets to break on the muddy pool edge, its large portly form, dazzling white plumage and long bent neck giving just a slight hint of menace as it gets nearer. This is the 'cob', or male, *mute swan* just coming to see what you are about, for you guess correctly that in the tall reeds not far away it has its nest with the female, called the 'pen', sitting dutifully on her eggs. As you rise to take a look, the cob grunts and hisses; mute it might be named because it has no call or song, but if you venture near its nest or young it will soon find its voice. You see the nest from afar, a great heap of reeds, rushes and rubbish of all kinds on which the great white mother bird is sitting, serenely confident that her mate will keep all intruders at bay, with a fierce determination and built-in strength.

The pen will lay 3 to 12, but usually 7, pale green eggs from March to May. The cygnets hatch out in 35 days and are soon in the water paddling behind their devoted parents. They grow fast, feeding with their parents on water weeds and other vegetation, but towards the end of the year they are firmly driven off by their parents to lead an independent existence.

Swans pair for life or until some disaster overtakes one or the other, and often nest year after year in the same site amongst the reeds. The mute swan was introduced into Britain by Richard I in the late twelfth century and it has virtually no foes. Indeed they are protected birds and those not marked as having a specific ownership belong to the Queen. What a wonderful sight to see a herd of swans in the air flying across the evening sky, the measured beat of their great wings making a wonderful wild melody, totally in harmony with the atmosphere of the island.

There is an apt description of these magnificent and striking birds by W.B. Yeats, that great Irishman, in his poem 'The Wild Swans at Coole':

But now they drift on the still water
Mysterious, beautiful;
Among what rushes will they build,
By what lake's edge or pool
Delight men's eyes when I awake someday
To find they have flown away?

The *coot* is not difficult to spot, with its jet-black head and body and conspicuous white frontal shield and bill. It spends most of its time on the water, diving for the succulent stems of water plants and for aquatic life such as water snails, insects and their larvae. Coots are territorial birds and in early spring they pair up and settle their territory; they will protect its integrity, voicing a hard, explosive and metallic sounding 'skik' at any other coot that encroaches. When this fails, they will engage in angry combat; watch them as they lie on their backs and try to grab an intruder's front with their feet.

The nest of the coot consists of a platform of decaying water plants on some support, like a sunken bough, or amongst aquatic vegetation such as reeds and rushes. The seven to ten stone-coloured eggs heavily speckled with dark brown spots are laid from April to July. Incubation takes about three weeks and is shared by both adults, and during this time the bird not sitting occupies itself collecting small sticks of reeds to constantly improve the nest. Watch the birds as one offers its sitting mate the material as a sign of belonging and affection and this bird then places them on the side of the nest, carefully pushing them into place. When the chicks hatch out as little balls of coal-black down with bright red heads, they are soon able to swim, and at any sign of alarm they tumble out of the nest into the water. They are fed small pieces of waterweed by their parents, and after a few days they lose their red colour and become somewhat dull and nondescript. The coot is often content to raise just one brood per year, especially if the first has been largely successful.

Coots are gregarious and sociable outside the breeding season and in winter they will seek out the company of other ducks, as well as their own species. Resident coots are joined by many from the mainland in winter and gather in quite large flocks on the open waters of Great Pool and Abbey Pool.

Look out for the smaller and more solitary bird, the *moorhen*, during your walk around 'The Pools'. The moorhen has conspicuous white stripes along its flank and a very distinctive red bill and forehead. A poor flier, the moorhen is truly at home on the water,

71

where it spends the whole year near thick undergrowth in which it can hide if danger threatens. It can remain submerged for quite long periods with just its bill above water. When disturbed on the water, it rises up with a pattering of feet along the surface, uttering its strong call, 'kittok'. Moorhens generally build their nest of reeds, rushes and grass amongst the plants that grow beside the water. It is less conspicuous than the coot and their 7 to 12 eggs are more buff-coloured and the spots are larger and browner than those of the coot. The breeding cycle of the moorhen is similar to that of the coot but it is more likely to raise second and even third broods. Like the coot, the moorhen feeds on just about anything it can find, from slugs and worms to seeds and waterweed; interestingly, it frequently devours the eggshells after the eggs have hatched, to retain the necessary calcium and mineral salts in its body.

One of the lovely sights of early summer is to see a *common shelduck* with its brood of seven or eight young black and white ducklings in convoy across the lake. This species nest regularly on Samson and St Agnes, and for some reason they risk the hazardous journey across the open sea with their young to rear them on 'The Pools' of Tresco. Perhaps they find more food and shelter here amongst the shallow placid waters of the island.

On your way to explore the coastline and cliffs to the north of Old Grimsby, you walk up through the woods and shrubs above Great Pool; a thrush breaks into a loud song, filling the grassy sunlit glade with music of amazing quality. This is the *song thrush*, which is much more common in Scilly than on the mainland, where over the past 25 years its numbers have declined by over 70 per cent. On Scilly the population is stable and is some 15 times greater than anywhere else in Britain, probably because of the warmer winters and possibly because slug pellets are seldom if ever used and their poison does not enter the food chain. The song thrush suffers badly in severe weather because of the difficulty in finding its staple food of earthworms and snails. Unique amongst British birds, this delightful songster has the habit of carrying snails to a particular stone and pounding them on it, thus shattering their shells. This stone is the bird's 'anvil' and you will find many surrounded by broken shells beside island paths. The song thrush augments its diet with berries and insects and, unlike many other birds, its voice can be heard throughout the winter on mild days, gaining strength and joyousness as the spring approaches and then trailing off in late June and July. It makes its nest in an enormous range of sites, but in February and March it usually nests 2 metres or so off the ground in a deciduous bush or hedge where it is well hidden. The nest of grasses and mud is lined with mud, unlike that of the *blackbird*, which is similar but lined with grass. The eggs of the song thrush are light blue with black spots and are laid as early as February, with repeat clutches on to as late as August. On the islands you will find the song thrush very tame, and at times when walking the coastal paths I have had a bird almost feeding out of my hand, taking pieces of bread from my sandwich with little apparent fear. This tameness is apparent in many of the birds in Scilly for reasons I just do not know; maybe it is in response to the friendliness and kindness of the inhabitants.

On the north-west side of a little boulder-strewn cove called Gimble Porth is a 10-metre-high cliff on which a colony of about 20 to 30 pairs of *kittiwakes* have made their home. These are possibly the same group that only a few years ago resided on the eastern cliffs of St Martin's, but there again they could have migrated from Men-a-vaur which remains the chief stronghold of the kittiwake. This delightful gull gets its name from the loud plaintive call of 'kitti-wa-ak' or 'kaka-week' which it utters around its breeding stations.

Some 40 centimetres in length, the kittiwake is not unlike the *common gull* in appearance, with grey wings and pure white plumage. However, it can be distinguished from the latter species by its black wing-tips and its slimmer and more dove-like look. With your binoculars, look at its legs, which are black rather than the greenish-yellow of the common gull. The common gull is a winter visitor and migrant on Scilly and does not breed, whereas the kittiwake is a regular breeder on the islands.

Like the fulmar and storm petrel, the kittiwake is a truly oceanic species, spending much of its time amongst the waves of the deep Atlantic, coming only to land in February and March to breed. What an attractive bird to watch at this time as they pair up, standing close together, breasts inwards and touching the cliff face and each other. Often they break out into soft mewing cries which spread to the whole colony. Then suddenly all is silent, and only the gentle wash of the sea on the rocks below can be heard. Their courtship exchanges are delightful; graceful bowing of heads to each other, and gentle fondling movements with their bills, all accompanied by excited mewing cries.

Soon nest building begins; the birds form collecting parties to tear grass and seaweed from nearby cliffs and rocks and cement this together with mud and much tramping of feet on the rocky ledges. In this way they form fairly secure nests, but well separated from each other so that each have a certain degree of independence to rear their young. The two or three eggs vary greatly in colour, but are generally greyish-white and heavily blotched with reddish-brown, and are laid in May and June. Incubation is by both parents and the young hatch out after about 22 days. In contrast to the savage and cannibalistic behaviour of some of the larger gulls, the kittiwake never preys on the eggs or young of its own kind or that of any other species and is truly an inoffensive and likeable bird. It feeds off small fish, plankton and crustaceans taken from on or near the surface of the sea. The young are fed on this food for about five or six weeks, when they leave the nest and fly out to sea.

By the middle of August the breeding ledges are empty and the presence of the kittiwake colony is marked by the white splashes of their excrement below their nests. Hopefully they will be back again to the same place next year, but who knows; they may move to a different location or join up with one of the many other colonies on the islands. Whatever happens, you will almost certainly see these gentle and loving birds again. Meanwhile, wish them well as they leave the friendly shores of Scilly to ride out the autumn and winter storms amidst the flashing foam of the mid-Atlantic.

Tresco, with its wide variety of landscape and geology, has a rich flora with well over 460 recorded species. Starting at Old Grimsby, let us walk around the island and see what we can find; June is a good month to visit as most of the species are then in flower. A patch of waste ground near the old stone jetty contains a number of different plants, amongst them *yarrow, opium poppy* and *long-headed poppy*.

From 15 to 30 centimetres tall, yarrow is a handsome plant with finely divided grey-green leaves and numerous daisy-like white and yellow flowers arranged in clusters at the top of the stem. On some of the islands, like Tean and Great Ganilly, the flowers are a deep purplish-rose colour, giving the whole plant a very striking appearance. It blooms from June to August and its flat looking flower-heads give off a strong aromatic scent. Yarrow is a common plant all over Europe and its Latin name, *Achillea*, comes from Achilles, who, when wounded at the siege of Troy, was told by the weeping Aphrodite to cover his wound with yarrow to ease the pain. Since ancient times this plant has been used to heal wounds, particularly wounds of war in battle and was once known as *knight's milfoil*. Yarrow was regarded as a powerful herb as far back as Anglo-Saxon times and yarrow tea is still drunk as a tonic and stimulant. On the Orkney Islands milfoil tea is drunk to dispel melancholy.

74

The opium poppy is a rare plant in Scilly but I have seldom failed to find it growing on waste places around Old Grimsby or elsewhere on Tresco. This poppy grows some 30 to 100 centimetres tall, has dull greyish-green lobed leaves and carries large bowl-shaped lilac or pink flowers at the end of erect, almost hairless stems. The opium poppy is not a natural British plant but originates from south-west Asia. It is the species whose milky sap is used for the production of opium but it produces very little in the cooler climate of the British Isles.

Nearby is another poppy, the long-headed poppy, which is common in the bulb fields and other cultivated ground in Scilly. This poppy differs from the *common poppy*, which is rare on the islands, by having a long, smooth club-shaped seed capsule at the centre of the flower head, rather than the spherical-shaped one of the latter. In addition, flowers of the long-headed poppy are sometimes paler in colour and do not have the dark centre often found in those of the common poppy; both varieties flower from June to September.

As you start your walk on the coastal path just to the west of the Blockhouse, look across the fields to the square tower of St Nicholas Church, built in 1879. In these fields covered with the yellow flowers of *smooth hawksbeard*, look for the striking white flowers of *wild carrot* and the less conspicuous but nevertheless intriguing *Babington's leek*.

The wild carrot is believed to be the plant from which the various kinds of table carrot have been evolved under cultivation. Normally it grows to about 70 centimetres tall, but here beside the church it is enormous, over 90 centimetres in height. Its roots, although having the same pale orange colour and smell of the cultivated variety, are much thinner and are not edible. From this rootstock arises the tough, branched stem which is ridged and densely covered with white hairs. It has feathery leaves and the white flowers, which bloom from May to August, are gathered together in large hemispherical clusters which are quite distinctive. The leaves, like the roots, give off a distinctive carrot smell when crushed.

Babington's leek is a common plant on all the inhabited islands but on the mainland it is rare and is confined to the coasts of the south-west. About 100 to 200 centimetres tall, this plant looks like a cultivated leek gone to seed with broad lance-shaped leaves, but its flower head, on a long stem, consists of many large purple bulbils, beside which sprout the small pinkish-purple flowers. You will find it in flower in July and August.

The coastal path now is flanked with tall *marram grass* and in places where it is well cropped by rabbits and kept short by walkers, it is covered in yellow *bird's-foot trefoil* and *lesser hawkbit*. From the Blockhouse all the way round the coast to Appletree Bay is an area known as coastal dune and blown sand. Here the sand has blown up from the beach, forming large dunes kept stable by the long trailing roots of the marram grass. Inland the sand has built up over the granite base to a considerable depth. This is a botanist's paradise and you will soon come across such plants as *haresfoot clover, common centaury, sea pearlwort* and *lady's bedstraw*.

The haresfoot clover is the easiest of all the clovers to identify because its flower heads are long, narrow and fluffy, looking just like the feet of a hare or rabbit; the fluff is made of long pinkish hairs which surround the small white flowers. The plant itself is up to 40 centimetres tall, with hairy stems and leaflets; the flowers appear in June and can still be found as late as October.

The common centaury is easily spotted because of its beautiful clusters of funnel-shaped pink flowers and its neat appearance. The plant is not more than 30 centimetres tall and has an erect square stem which rises from a rosette of oblong leaves. It blooms from May until September. The centaury is a valued herb which is collected and dried and taken in tea form to aid digestion and relieve flatulence and heartburn. That seventeenth-century astrologer-physician Nicholas Culpeper said that the plant 'cleared the eyes from dimness and cured adder bites' – but there is no need to use it in Scilly for the latter!

The diminutive white flowers of the sea pearlwort are much more difficult to find than the previous species; but it is abundant on all the islands, growing just about anywhere on bare places on the dunes, on cliffs and in rocks and walls. There are many other common species of pearlwort in Scilly, all of which have awl-shaped leaves joined in pairs along the stem. The sea pearlwort can be distinguished from the others by its fleshy broader leaves and by the way the sepals (small leaves around the petals) almost completely enclose the seed capsule. The plant is in flower from May to September and you will need your magnifying lens to distinguish its features.

Adding colour to the sandy places beside the coastal path from June to September is the lady's bedstraw. Its stem is some 30 centimetres long and bears whorls of some 8 to 12 slender leaves along its length. At intervals, clusters of small cruciform-shaped bright yellow flowers are borne on short stems. The plant forms quite large colonies in places where the foliage straggles through the grasses and these are easily seen and recognised. This plant, when dried, gives off a pleasant scent of new-mown hay and many years ago was included in straw mattresses, especially in the beds of women about to give birth; hence its name. The flowers were once used to curdle milk for the making of cheese, and it has also been used as a dye.

The sun is gaining strength and it is time to take a break and admire the view from Lizard Point, so named because it is on the same latitude as the Lizard Peninsula in Cornwall. The sea is calm, changing colour in a subtle way, dark blue to sapphire and now pale turquoise as the sun breaks out behind a small patch of grey stratus cloud. To the north the light on Round Island still continues to flash regularly, and in the foreground some *oyster-catchers* are feeding on the shore, calling out their long piping trill. They fly off with strong shallow wing-beats, disturbed by a couple of young children walking hand in hand along the beach talking excitedly and picking up the many kinds of shells that litter the sand here. A white and blue sailing boat, serene in this perfect setting, is moored just offshore. Look to the east at the Eastern Islands, little humps of green on the far horizon, and as you turn your head around further, the long white beach of Pentle Bay dominates your view. You can hear the repetitive call, 'coo-eep, coo-eep' of the ringed plover, but try as you might you cannot see the bird. Beyond lies the island of St Mary's with its tree-clad hills, green fields and the comparative hustle and bustle of a town with cars and shops.

As you make your way by a narrow path through the marram grass and patches of bracken to Abbey Pool, a patch of yellow 'button daisies' growing on a bank of blown sand catches your eye; you look around and see another patch and yet another. This is *chrysocoma*, a native plant of South Africa that has become naturalised on the dunes of Tresco, having spread by wind disper-sal of its seed from the Abbey Gardens. You also see nearby, just coming into bloom, the deep violet-blue flowers of *agapanthus*, sometimes called 'the Kaffir Lily of South Africa'. These striking and most beautiful flowers are held in clusters at the top of thick 100-centimetre-tall stems; another escapee from the Abbey Gardens that has become naturalised on Tresco, St Mary's and St Agnes.

As you near the pool you see a number of *swallows* with their long tail feathers and red chins swooping fast and gracefully above the reflections of the grey Abbey walls in the water. They are feeding on the mass of flying insects, midges and small flies, that are about in the calm warm air. The tall spikes of *purple loosestrife* can be seen growing beside the water's edge. About a metre tall, this is a handsome wild flower with long narrow spikes of starry-shaped reddish purple flowers which are just beginning to open. It has narrow lance-shaped leaves which clasp the angled stem in pairs, and flowers from the end of June to August.

As you walk around the water's edge on the eastern side of Abbey Pool, see if you can find the diminutive plants of *bog pimpernel* and *sea milkwort* growing close to the ground in the grass just above the mud at the margin of the pool.

Bog pimpernel is one of our most beautiful plants and here you cannot fail to spot it as it makes a carpet of rosy-coloured flowers at the edge of the pool in June and July. The creeping stems root into the moist ground and the numerous pale oval leaves are contained beneath the much larger flowers which dominate the whole plant. Each of the five petals of the flower has delicate bright pink veins running through it, which add to their attractiveness.

In the damp grass, the sea milkwort is less easily seen as it is a small, slender plant which creeps close to the ground, its pinkish and white flowers being much less striking than that of the pimpernel. It has elliptical, stalkless leaves, which, like so many other shore-plants, are thick and fleshy, and it flowers from May to August.

A much taller plant found in the longer grass in damp places here and around Great Pool is the *creeping forget-me-not*. Flowering from April to July, this plant is some 60 centimetres tall, with its lower stems abundantly covered with hairs. The five-petalled flowers are a beautiful pale blue with yellow centre. Our forefathers knew this plant as the *creeping mouse-ear scorpion-grass*, because as the buds open in succession, so the stalk lengthens and curves in the shape of a scorpion's tail. The Latin name of the species, *Myosotis*, derives from two Greek words meaning 'mouse-ear', which refers to the hairiness and shape of the leaves. In the Middle Ages the stem of this plant was worn as a symbol of love by young people, and its original name was slowly replaced by the present-day 'forget-me-not'.

In wet places around Great Pool, you will almost certainly find a water plant of the buttercup or *Ranunculus* family, namely the *lesser spearwort*. As its name implies, its leaves are long and spear-shaped. It grows some 40 to 50 centimetres tall, with small yellow flowers, less than 2½ centimetres across, held on short stems, and it flowers from May to July. Like most of the buttercups, the stem exudes an acrid poisonous juice when crushed and the leaves, if eaten, have a burning taste; hence its Latin name *flammula*, a 'little flame'. 'Ranunculus' is derived from the Latin name *rana* or 'frog', an allusion to the damp meadows and ponds where most of the species are to be found.

Another attractive plant you should be able to find fairly easily in damp sandy places on Tresco is *yellow bartsia*. A scarce species on the mainland and confined to the south and western coasts, it is very common on the inhabited islands of Scilly. Interestingly, yellow bartsia grows only where the mean January temperature remains above 5°C, suggesting that it requires a mild winter to survive; clearly this is why it is so prolific here. From 15 to 50 centimetres tall, this erect, unbranched, grey-green plant is covered with sticky hairs and has toothed, unstalked, lance-shaped leaves. The large yellow flowers, whose lower lip is much larger than the upper, bloom from May to September.

A very common and beautiful summer flower to be found just above the high-water line is the *sea bindweed*. A member of the convolvulus family, this plant sends runners through the sand and shingle and at intervals puts up glossy heart-shaped leaves and large, striking, funnel-shaped pink flowers with white stripes. The sea bindweed will only thrive where its roots can get a real taste of the sea and so in many places it actually grows on the sandy shore itself, especially on the beaches around Tobaccoman's Point. It is also found creeping amongst the marram grass on the dunes.

An unusual and uncommon plant which you could find growing just about anywhere on Tresco, especially beside the coastal path, is the *lesser broomrape*. This parasitic plant has an erect, scaly, pale reddish-purple stem up to 40 centimetres tall. The pale yellow and purple flowers are well spaced up the stem and can be found from May onwards to September. There is not a hint of green to be seen on this plant, indicating a complete absence of chlorophyll, that ingredient essential to the process of manufacturing cell material from the elements. So the plant must find sustenance in another way and it does this by attaching its roots to that of various other species and so stealing the food from them – thus the broomrape family are known as root-parasites.

This has been a brief account of just some of the many wild flowers to be found on Tresco; you will find many others, like *sea rocket*, *rock spurrey* and *Bermuda buttercup*, which have already been described in other chapters. Many of the flowers you will find are exotic and will have escaped from the tranquil seclusion of the

Abbey Gardens, a place where time stands still amongst the ruins of the ancient Abbey. This is yet another botanist's paradise, with specimens from as far away as South Africa, Australasia and the Chatham Islands in the Pacific, all collected by the Dorrien-Smith successors of Augustus Smith. Today they delight the many visitors who, even if they have little horticultural knowledge, can still enjoy and appreciate the true beauty of it all. The gardens also provide the resting places for ships' figureheads of vessels that foundered in Scillonian waters. Called Valhalla, this is a fitting description of the carved wooden sea gods and goddesses that once sailed the seven seas on ancient ships that met their fate on the rocks of Scilly. So enjoy not only the splendour of the flowers amongst the grasses and sand of Tresco, but take time to explore the world famous Abbey Gardens and its 6,000 varieties of tree, shrub and plant; you will not be disappointed.

Butterflies of many species abound on Tresco and when the sun is shining from a clear blue sky they will be all around you, flitting from flower to flower in the fields and woods and beside your path, adding to the ever-changing shades of colours, yellows against the browns, the greens against the white beaches and omni-present above all, the deep, deep blue of the sea and sky. *Common blue* butterflies, *meadow browns*, *peacocks* and *small tortoise-shells*, all described in previous chapters, will be here to delight you.

The bright red and black colours of a *burnet moth* suddenly catches your eye, brilliantly set off by the yellow of a *lesser hawkbit* flower. The burnet moth's bright colouring, with six red spots, acts as a warning to potential predators, since if attacked it will exude a yellow fluid which contains hydrogen cyanide (prussic acid). Birds who get a taste of this moth will not be tempted again! The burnet moth is not harmful to humans unless eaten, so be careful when picnicking that they do not land on your sandwich. There are seven species of burnet moth in Britain but the six-spot variety is the one you are more likely to come across in Scilly. The moths live in colonies and may be numerous in places where *vetches* and *trefoil* plants grow. The eggs of the female moth are laid in groups on these plants and hatch out into somewhat dumpy caterpillars which are various shades of pale yellow with numerous black marks; their speckled appearance camouflages them well on their food plant. The caterpillars hibernate in autumn and winter on the plant and resume feeding in the spring. In early summer they spin a cocoon around themselves in which they pupate in the form of a chrysalis. The yellow boat-shaped cocoon is attached high up on a grass or flower stem and the moth emerges in June and July, leaving behind the black chrysalis case half-hanging out of the cocoon. Enjoy the sight of these moths among the sandhills along the coast and admire their bright metallic colours and their slow easy flight across the grass.

The *great green bush cricket*, the largest of its species in Britain, breeds on Tresco and you may find it just about anywhere on patches of bramble or gorse. About 6 centimetres long, commonly, but mistakenly, called the great green grasshopper, it is in fact a bush cricket with its long wings and antennae and even longer powerful hind legs. Bush crickets eat both plants and soft-bodied insects and in turn they are preyed upon by birds and rodents; but their green colour, which camouflages them well in the thick vegetation, and their secretive habits, help them to survive. Only the male of the species produces the characteristic strong shrill 'chirrup' as it scrapes its left forewing across the edge of the right forewing. This burst of song helps it to attract females and to warn off rival males. Their eggs are laid singly on rotting wood or crevices in bark in the autumn and these hatch out into young bush crickets in April without going through a chrysalis stage.

You may find *common field grasshoppers* amongst the summer grasses and these have much shorter antennae than the bush crickets. Their colour is very variable, ranging from green through buff to purple, and this helps them to blend in with various types of vegetation and background. They live for about five months on a diet of grass, and their eggs are laid in batches of up to 14 in a foam-like secretion in the soil from which the young hatch in the spring.

Autumn arrives almost imperceptibly when, with the alchemy of the season, the colours of nature change so cleverly; the orange and red fruits of berries appear on wayside shrubs, the leaves of the woods change to subtle yellows and orange and the reeds around 'the Pools' are touched to brownish-white, their feathery flower spikes making soft rustling in the wind. It is time for fungi to make their appearance as the October mists sweep in from the sea, giving them the moisture they require for their mycelium to begin to work.

The *parasol mushroom* will often be seen amongst the grass at the edge of pathways and copses. This is one of the largest British fungi, up to 40 centimetres tall, with a greyish-brown cap some 20 centimetres across. The cap at first is shaped rather like a drumstick but later it splits away from the stem and expands into a parasol and is covered with dull brown shaggy scales. The point where the cap is joined on to the stem is marked by a double ring of tissue. I have found this fungus growing as early as June on the path just below the Blockhouse. The parasol mushroom is excellent to eat, with a sweet taste when fried in oil, although the stem is somewhat tough.

Look out, too, for the *scarlet hood*, which is one of the most striking fungi of the grass-covered sandy wastes, especially where the turf is short. As its name suggests, it has a bright scarlet bell-shaped cap some 2 to 4 centimetres across and is said to be good to eat but I have never tried it; to me, the colour is off-putting!

Before leaving Tresco take one last stroll along the southern sandy beaches, some of the finest in the world, and see how many different shells you can find. Look particularly for *guinea money* shells of the cowrie species with their beautiful curved shape and delicate pink lines curving over the top. Known also as *European cowries*, these shells, less than a centimetre long, belong to the univalve molluscs that live inside them when they are in the deep ocean. You will more likely see them washed up and empty on the sand just below and above the high-tide mark.

I have really only touched on what this wonderful island, this Elysian Field so varied in its landscape and scenery, has to offer, and much, much more could have been written about such places as Piper's Hole, that fascinating subterranean passage just north of Gimble Porth, or about The Abbey, Cromwell's Castle or Castle Down, to name but a few. Do make use of the many guidebooks on Scilly for they will give much information about these places.

Appropriately, our point of departure is New Grimsby, the capital of Tresco, with its appealing harbour and rows of old-world granite cottages. From the quay look across New Grimsby Channel and see the small island of Bryher only half a kilometre away.

Ringed Plover

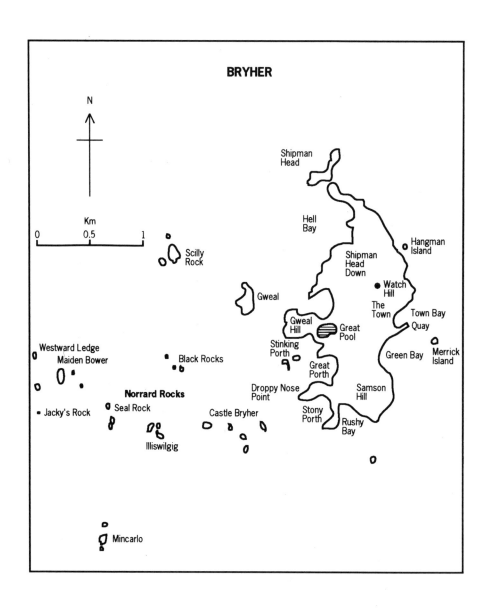

BRYHER

N

Km
0   0.5   1

Shipman Head

Hell Bay

Scilly Rock

Gweal

Shipman Head Down

Hangman Island

Watch Hill

The Town

Town Bay

Gweal Hill

Great Pool

Quay

Stinking Porth

Westward Ledge
Maiden Bower

Black Rocks

Great Porth

Green Bay

Merrick Island

Norrard Rocks

Droppy Nose Point

Samson Hill

Jacky's Rock

Seal Rock

Castle Bryher

Stony Porth

Rushy Bay

Illiswilgig

Mincarlo

7

BRYHER

The Quay

Bryher is an island of interesting contrasts which you will enjoy as you explore its wild, harsh and rugged coastline and feel the embrace of its pastoral landscape. To the north-west, gentle hills sweep down to a desolate scene. Here, black granite rocks alone provide a bulwark to the winds and seas that charge in from the great ocean to the west and end up as a swirling mass of foam and spray crashing up and over them like demented devils; this place is not called Hell Bay for nothing. In contrast, Rushy Bay in the south has a gem of a beach around which wild flowers grow in abundance and where the only intrusion on your peace and solitude will be the soft lapping waves, the gulls' call and perhaps the twitter of a flock of *goldfinches* as they gather in the autumn to feed on thistle heads.

A small island less than 3 kilometres from north to south and not more than 2 kilometres wide, Bryher is the least populated of the inhabited islands, with just some 76 inhabitants, who gain a somewhat precarious living from horticulture, fishing and tourism.

Numerous graves and field systems are scattered about Bryher and these, together with the Iron Age cliff castle at Shipman Head, suggest that it was well populated in ancient times, especially on the more sheltered eastern side. The Bronze Age graves and cairns can be seen prominently on Gweal Hill, Samson Hill and Shipman Head Down, together with low stone and earth banks indicating prehistoric field walls. The remains of ancient roundhouses and field boundaries are also evident within the intertidal zones of Great Porth, Green Bay and Town Bay. So history and the spirits of Bryher people are all around you as you walk about this romantic island of hill, rock, heather and cultivated fields.

As you walk up the steep path to Shipman Head, look down at Hangman Island, a conical lump of rock rising sheer out of the water with a mock iron gallows on the top. Did it get its name from the Civil War when defiant Royalist soldiers were executed here by Parliamentary invaders, or does its origin go back much farther, to the days when pirates roamed the seas and it was common for them to be hanged when caught?

The views are breathtaking as you near the rocky crags on the top of Shipman Head Down. To the west are the cold, black Norrard or 'Northern' Rocks, sticking out discordantly from the rolling blue seas. How descriptive and evocative are their individual names, Seal Rock, Westward Ledge, Maiden Bower, Jacky's Rock and Illiswilgig, to name but a few. Looking at those unfriendly and remote spots of land, it is difficult to imagine how the *wren* manages to breed on some of them, like Castle Bryher, but it does. Birds are all about you on the cliffs, *black-backed gulls* and *herring gulls* make their presence known with their strident calls competing with the noise of crashing waves; and *shags* stand with wings outstretched on their communal roosts. In contrast as you look to the east there lies the friendly soft lines and peaceful serenity of Tresco.

Mincarlo, the largest of the Northern Rocks, is an important breeding site for birds as well as of the *Atlantic grey seal*. Here are colonies of *razorbills, guillemots, puffins, shags* and probably *storm petrels*, while a few *cormorants* build their nests on the rocky ledges and among boulders. Mincarlo, like many of the other islets, is exposed to the full force of the Atlantic and south-westerly gales, and at times the sea breaks right over the island; plants that grow here must be firmly rooted, and so you will find *tree mallow, thrift, sea beet, common scurvy-grass* and even a few pink flowers of the *rock spurrey*. A very difficult island to land on except in calm weather, nevertheless it is a place to view from a boat gently riding the swell a few metres offshore. The jagged granite grey rocks stick out of the leaden blue ocean, while black and white razorbills and guillemots ride effortlessly over the undulating waves.

In March and April some of the hills turn to a delicate shade of yellow as the *gorse* comes into full bloom. On top of a gorse bush a chestnut-brown bird sits singing a varied soft and low musical twitter mixed with some sweet and shrill notes. This is a male *linnet*, about 13 centimetres long, with a pinkish breast which in the spring and early summer turns crimson. The female linnet lacks the pink and crimson colours and has a browner back with an overall more brownish streaked appearance. In flight, linnets can easily be recognised by their undulating flight pattern and their rapid twittering calls. Their nest is built of roots, twigs and grass, lined with hair and feathers, at just about any height up to 4 metres, in my experience nearly always in a gorse bush; the four to six whitish eggs, speckled with red and brown, are laid from April to June. Outside the breeding season the linnet, like most finches, is gregarious and is often seen in large flocks, feeding on weed seeds.

As you walk southwards, leaving the exposed headland with its wind-pruned vegetation of maritime grassland and waved heath, you come across Great Pool, a circle of brackish water which, not surprisingly, is often polluted by the sea breaking over the high rocky bank and sweeping flood water into the surrounding low ground. *Grey mullet* find their way into the pool from the sea and feed on the microscopic organisms in the mud and soft sand and on the green algae on the rocks. The mullet has no teeth, but skims the pool bottom with the fine comb-like fringes of its thick lips, retaining the mud and the small animals it contains, while puffing out the larger unwanted particles. The grey mullet has a grey-green back and white belly and has two top fins, which often break the surface of the water. The fish return to the sea after breeding in the pool, probably down the small rivulet that flows south into Great Porth.

The area around the pool and in the shrubs and trees to the east is as good a place as any to watch the birds during their spring and autumn migrations. Scilly is renowned as a place for rare birds, and the numbers seen are greater in the autumn months from August to early November than in the spring peak of mid-March to early May. This is because in the autumn many of the successfully reared young accompany their parents on their return journey and the birds have time to linger in the warmth of the islands, whereas in the spring they are in a hurry to reach the mainland to breed, staying just long enough to feed on the plentiful food before moving on. Irregular migrants such as the *hoopoe* and *golden oriole* may be seen, as well as extreme rarities such as the *Caspian plover* and *Pallas's warbler*. On the other hand, many of our common migrants, such as *swallows, willow warblers* and *flycatchers*, are seen every year, sometimes in great numbers. For many thousands of years our birds and their ancestors have undertaken their many and varied journeys over sea, land and desert. Clearly, breeding and food motivates them to a great extent, but nobody really knows why so many birds migrate. What great mystery of evolution compels a bird like the *Arctic tern* to fly halfway around the world and back in a year? How do the birds navigate over the great oceans, at times through fog and great storms?

Perhaps one explanation could be associated with the phenomenon of continental drift. Some 300 to 400 million years ago the continents were joined together in one great land mass. Gradually this land mass split apart and huge pieces drifted away from each other on molten plates deep within the earth's crust, so forming continents. Over the past 200 million years the continents slowly established their present positions, although even now they are still shifting relative to one another. At one time, then, the ancestors of our birds, the reptiles, lived in one place, relatively speaking, and had no need to travel great distances for food. Imperceptibly, over the aeons of time, the first flying animals (birds) of the early cretaceous period 120 million years ago needed to travel further and further to get to their food and this distance got greater and greater as the land masses drifted apart. Their breeding cycle, too, would have been affected, and so over many millenniums of time birds gradually evolved the skills necessary to navigate the increasing distances required to meet their needs. Whatever the reason, the birds give us a marvellous spectacle as you see them congregating sometimes in large numbers, ready to leave on their long flights.

Scilly is also a very important area for wintering birds, which arrive in this relatively warm haven from the mainland whenever harsh weather occurs there. All they require is an adequate food supply, clean beaches and minimum disturbance as they alight on the islands, happy to be clear of the frost-hardened ground of the mainland. Some of the early arrivals are the *redwings* and *fieldfares*, both members of the thrush family, which can sometimes be numbered in their thousands. The redwing can be recognised by its white eye-stripe and chestnut-red flanks; while the fieldfare is larger, has a grey head and rump and a rusty-coloured back. Both these birds migrate south from their breeding grounds in the forests of Scandinavia and join others, like the *lapwing* and *golden plover*, which have not travelled as far. Redwings and fieldfares can both be found in mixed groups in the cultivated fields, looking for insects and worms amongst the vegetable remains. Often one bird acts as a lookout, sitting as high as possible on a post, tree or bush. Try to spot this bird before it gives a harsh chattering alarm call which sets all the birds to flight.

The islands also provide a winter refuge for birds, like the *chiff-chaff*, *black redstart* and *blackcap*, which are generally regarded as summer migrants to the mainland. These birds spend most of their time feeding on the prolific insect life of the weed-strewn beaches rather than concealed in the woods and bushes of their usual habitat. *Swallows* too sometimes find that the frost-free winters and the abundant insect life enable them to overwinter rather than attempt the dangers of a long migration south.

Just to the west of the Great Pool is Stinking Porth, where the granite boulders scattered all around the area are covered with greyish-green lichens, some 5 centimetres long in places. Just south of this bay is Great Porth, and between the two you will see one of the old gig sheds, now used as an artist's studio, which housed those important boats so often used for rescue. Imagine a foggy night nearly half a century ago in 1955 when the repeated blasts of a ship's siren were heard all over the islands. This was the *Mando*, a 7,176-ton freighter making its way cautiously north of Bryher, bound for Rotterdam with a cargo of coal. Suddenly the siren sounded continuously and the islanders knew instinctively that the ship was in trouble. The island's men rushed for the gig, which they quickly manned and rowed through the treacherous rocks and thick fog around the west coast. The *Mando* had run aground on the Golden Ball Brow near St Helen's but the St Mary's lifeboat, *Cunard*, had already rescued the 25 crew of the freighter and had both of its lifeboats in tow. The gig met the *Cunard* off Shipman Head, and while the rescued crew were taken on to St Mary's, the gig took the lifeboats to Bryher. Just another incident of rescue at sea that over time has become so commonplace to the hardy Scillonians. One interesting sideline is that the *Mando's* Italian chef had been wrecked before in Scilly, in 1926 on board the SS *Isabo*, and this time the same lifeboat coxswain rescued him again, nearly 30 years later.

As you reach the southern extremity of this most delightful of islands, you come across a beach of wild flowers called Rushy Bay. This place got its name from the fact that in the 1830s Augustus Smith quickly realised the potential dangers of coastal erosion on the sandy stretches of beach, so he had *marram grass* (rushes) and later *Hottentot fig* planted, and it was these with their long rooting stems that did so much to stabilise the whole dune system.

Within the dune system around Rushy Bay you will find little deep hollows where you can lie on the soft silvery sand with the sun on your face and the wind blowing above you through the grasses. Here are little hideaways where you can rest undisturbed. You look straight up at the azure blue depth of the sky and watch the small cottony-white clouds pass across, blown by the wind. Your mind turns to the stars that you know are there but are hidden by the bright light of the sun. You wish you could reach up and touch them to see what they are about but let them be to drift through space and to show themselves again when the earth rolls onward into night. You recall the lines of Longfellow in 'Evangeline' when he wrote:

> Silently one by one in the infinite meadows of heaven
> Blossomed the lovely stars, the forget-me-nots of the angels.

Your mind begins to wander as a deep peace takes over your body. A harsh croaking awakes you from your reverie. You raise your head and look around. In front are the grey granite rocks and the sea, and behind you are the tall hedges of *pittosporum* and *euonymus*, giving shelter to the narrow rectangular-shaped bulb fields. Then you see the source of the noise, a large *grey heron* standing motionless beside a rock pool, its long dagger-shaped bill ready to strike. Suddenly like lightning it stabs downwards into the water and comes up with a small speckled grey *scad fish*. As the ripples form larger and larger circles in the still dark pool, the heron takes to the air with slow and powerful deep wing-beats to find another quiet place and potential victim. The grey heron is a regular migrant and winter visitor to the islands and often gathers in quite large numbers on the rocks; it can be seen at any time of the year but it does not breed in Scilly.

Look at those two snow-white heron-like birds standing on the far rocks. With your binoculars you see that they have a slender, straight black bill with black legs and bright yellow feet. These are *little egrets*, which used to be uncommon migrants from southern Europe, but in recent years they have been seen in increasing numbers around the islands. I believe that global warming has had its effect here, and it may not be long before this bird begins to breed and nest amongst the bushes and small trees of Bryher and nearby Tresco.

If you walk across the rocks at very low tides, look into some of the large deep pools and admire the wonderful sea life thriving in the clear, clean water. Many kinds of *sea anemones* can be found attached to the rocks, each with its crown of stinging tentacles which move about in the water, trying to entrap shrimps, worms and other small sea creatures. The most common sea anemone to be found is the *beadlet*, about 3 centimetres tall, red, green or brown with a ring of 24 small blue spots on the top. When exposed at low tide it contracts into a mass of rounded, flat-topped stiff 'jelly'. The main enemy of the anemones is the *sea spider*, which clamps itself onto the soft parts and then proceeds to devour them, apparently immune to the stinging cells. The sea spider, so named because of its eight legs, is a dirty yellowy-grey animal with a body only about 2 centimetres long. The female ejects eggs from her legs; they are fertilised by the male, who attaches them with a type of glue to a pair of smaller legs which he possesses in addition to the normal eight. He then broods them until they hatch. The sea spider can often be found underneath small stones, but if you look for them be sure to replace the stones carefully in the same position as before, thus ensuring minimum disturbance of this small but important habitat.

Look for the tell-tale presence of a *common lobster* with its bright orange-red feelers just showing from under a large boulder. If you are very careful you can tempt this great predator of the pool out from his lair by gently scratching the side of the rock with a long thin stick. The lobster comes out to investigate whether this is another creature that it can eat by crushing it with its huge strong claws; slowly emerging, it shows off its beautiful brown and blue markings. Lobsters can live for 30 years or more, and as they grow they shed their old skins, growing new and larger ones. A lovely creature which is worth all the patience you will need to find it.

Bryher has a wonderful variety of wild flowers. As early as February you will come across the rosy-purple flowers of the *musk storksbill*, a plant that is a scarce species on the mainland, but is abundant in cultivated land and beside tracks on all the inhabited islands. Unlike the sea storksbill and common storksbill, both described in Chapter 2 and both found on Bryher, the musk storksbill is much larger and has a strong smell of musk.

Another plant of the bulb fields and sandy banks which is rare on the mainland but locally common on the inhabited islands is *toothed medick*. A member of the clover family, this plant has leaves divided into three similar leaflets and bears small pea-shaped yellow flowers at the end of long stems. Similar to toothed medick, but larger in all respects and often with black spots in the centre of the leaflets, is the *spotted medick*. It can also be distinguished by the expert from the former species by an examination of the small leaf-like appendages located at the point where the leaf stem joins the main stem. These stipules, as they are called, are deeply toothed in the toothed medick, whereas in the spotted medick the indentations are far shallower. The spotted medick is very common on the side of paths and in gardens and you will have no difficulty in finding it. The toothed medick is less obvious; look for it around the sandy wastes of Town Bay.

The real gem you may find on Bryher is the *spring squill*, an attractive plant of the lily family. Only 6 to 20 centimetres high, the spring squill has three to six narrow, shiny, often twisted leaves which arise from a bulb. The leafless stalks bear short spikes of star-like violet-blue flowers which bloom in late April to May. The plants congregate together and form matted areas on the short turf on cliff slopes, especially on Samson Hill and Shipman Head Down. You may also find it on the north-east side of Gweal Hill. If you fail to discover it on Bryher you will almost certainly see it beside the coastal path just north of Pelistry Bay on St Mary's.

On the heathy ground on the hillsides and downs, usually in amongst short grass, look out for the *lousewort* and *eyebright*, both semi-parasitic plants that attach their roots to those of grasses and heathers to rob them of sustenance.

Lousewort thrives on poor soil lacking in nutrients and was so named because our forefathers thought it produced the lice which infected sheep. However, the only connection between the plant and lice was the poor pasture on which both thrived; sheep were often grazed on poor land and the resultant lice infestation was incorrectly blamed on the plant. Lousewort straggles low on the ground and has thick, deeply cut glossy leaves, and eye-catching pink snout-ended flowers which bloom continuously from April to September. On the mainland this plant likes damp ground, but here in Scilly it seems to prefer drier places.

The bright flowers of eyebright seem to peep out at you from the grasses throughout the summer months; it has deeply cut, dark green oval leaves which grow in pairs up the central stem. The flowers bloom from April to September, and if you examine them with a hand lens you will see how attractive they are, white with beautiful markings of yellow and purple. There are many different varieties of eyebright, and their Latin name is derived from the Greek word *euphraisio*, meaning to delight or gladden, an allusion to the pleasing colours of their flowers or to the joy of having eye complaints cured with their extracts. Herbalists in the Middle Ages prepared a powder from this plant which was used for brightening the eyes, and in medieval times potions made from it were used to cure short-sightedness and dimness of vision. That lovely bird the *linnet*, mentioned earlier, is fond of pecking at eyebright but I cannot say how you would recognise a short-sighted linnet!

On the sandy soils around Green Bay and in the cultivated fields, you may be able to find a diminutive plant with small pale yellow flowers which sprawls across the ground with trailing stems up to 30 centimetres long. This is the aptly named *small-flowered buttercup*; in bloom from April to June, it can be easily identified by its furrowed and hairy stalks. These support the flowers, whose tiny petals are dwarfed by the bristle-covered immature fruits. Another similar plant, the *prickly-fruited buttercup*, grows commonly in similar places; this buttercup has larger yellow flowers and can be easily recognised by its prickly fruits. It is in flower from as early as February but dies off in May or June. Both these species are Mediterranean in origin, but whereas the small-flowered buttercup is a true native of Scilly, the same cannot be said of the prickly-fruited buttercup, which has become naturalised only in the present century, possibly having been brought in with imported grain. Wherever its country of origin, it is now well established on all the main islands and is considered to be a 'bulb field pest'.

On the shallow granite soils and amongst the short grazed *ling* heather on Samson Hill and on the south-east of Shipman Head Down grows another speciality of Scilly, and known elsewhere in Britain only in the Channel Islands; this is the *orange bird's-foot*. Like the *common bird's-foot trefoil* mentioned in Chapter 2, this inconspicuous member of the pea family trails across the ground, but unlike the former it has orange-yellow flowers and has fewer leaflets. The orange bird's-foot flowers from May to September and grows locally on some of the other larger islands in Scilly like St Martin's, Tresco, St Agnes and Tean, so see if you can find it; you will enjoy the search and the elation when you do.

Frequently associated with this plant and growing in the same habitat is the much commoner *hairy bird's-foot trefoil*. This is similar to the common bird's-foot trefoil described in Chapter 5, but it is a smaller plant, has usually only two to four flowers in the flower head and is covered all over with hairs.

One of the lovely sights in May and June is to look at the golden blooms of *corn marigold* making a dazzling and delightful pattern of colours across the bulb fields, especially when seen from a high vantage point. On the mainland this plant has all been eradicated from cultivated fields by the use of modern weedkillers because it doesn't dry out easily and it tended to rot the straw bales, but in Scilly its beauty is really appreciated. It grows on all the inhabited islands and is more than holding its own. The near absence of cereal crops in Scilly has meant that there was no economic necessity to eradicate the corn marigold and so it has flourished, and indeed at one time attempts were made to market the flowers. It is truly a magnificent plant, up to 80 centimetres tall, with bluish-green fleshy leaves and large bright golden-yellow flowers some 3 to 5 centimetres across. The lower leaves are deeply toothed and stalked but the upper ones clasp the stem and are just slightly toothed. Let us hope that this striking weed of the arable fields will continue to thrive and give pleasure to everyone who sees it.

It is September and the evenings have started to draw in and one has that feeling that autumn is just beginning to nibble at the lingering edges of summer. This is a feeling sensed and shared by all those who, like myself, experience and love the aura and sheer loveliness of the changing seasons here on the islands. One of the late-flowering species that is worth searching for is the *autumn lady's-tresses* already described earlier in this book, so do look for it on the sandy turf around Great Pool.

Butterflies are plentiful on Bryher and one to look out for is the *painted lady*. This is a migrant butterfly that arrives in April or May, having flown over 1,000 kilometres from south-west Europe and North Africa. It is a powerful flyer and it seems to average some 10 to 15 kilometres per hour on its long journey. There is some doubt, however, whether it is the same individual that makes the whole journey, or whether it breeds en route and it is its progeny that completes the journey. In some years the number coming to Britain is huge; many tens of thousands in 1969, 1980 and 1996, but in other years they are scarce.

Painted lady butterflies breed in large numbers all over the Middle East; in Iran, Iraq and in Egypt; and huge migrations of them occur in Eastern Europe. The blood and carcass of this butterfly is dark red in colour and it has been suggested that it was the remains of millions of them that turned the rivers of Egypt red as described in the Old Testament (Exodus Chapter 5). With a wingspan of nearly 6 centimetres, the painted lady is one of our larger butterflies; it is easily recognisable with its striking colours of brown, red and black and with prominent white marks on the black tips of the forewings. When resting on bare ground with its wings closed, it blends in well with its surroundings, and it craftily points its head towards the sun so that no revealing shadow is cast to give away its position to a passing bird. The early migrants lay their eggs on a variety of wild flowers but mainly thistles, and they hatch out in about a week into black caterpillars with yellow spines and yellow stripes down their side. These give rise to a single generation of butterflies, which are seen in late summer and autumn. Unfortunately these soon die as colder weather arrives and, unlike some other species, such as the *small tortoiseshell* and *peacock*, they have not learnt the trick of hibernating.

Just occasionally on a hot summer day between June and September you will see what looks like a small hummingbird hovering with fast-beating wings in front of a honeysuckle or large garden flower. This is the *hummingbird hawk-moth*, another migrant from southern Europe and the Mediterranean. Brown and orange in colour and with a wingspan of some 6 centimetres, it is a fast flyer, often covering 140 kilometres a day. In some years it arrives in large numbers and is able to breed, laying its eggs on the *lady's bedstraw* plant in July and August. The new generation of moths either migrate back to Europe in late autumn or occasionally hibernate, and if the winter is particularly warm and they survive, then they will be seen again as early as February or March. Unlike most moths, this species is a day flier, with a characteristic darting action; watch it as it hovers in front of the flower and inserts its long tongue to find the nectar, then suddenly as quick as a flash it disappears, only to reappear a few seconds later.

Before leaving Bryher with its friendly people and its sense of quiet, settled permanence, take a walk along the lane through Bryher Town with its line of small houses and up to the top of Watch Hill. Here amongst the gorse and boulders is a small walled enclosure now used as a lookout for shipping; there was once a larger structure here, a coastguard hut which also served as a mark for island fishermen. The view is breathtaking; the Norrard rocks to the west surrounded as always with splashes of white water, and to the east the vista of Tresco, tranquil and untroubled, with the ever-flashing light of Round Island just showing above its trees. The contrasting topography of Bryher itself, with its northern heathland, its rounded hills, sheltered valleys and tiny bulb strips with parallel lines of shelter hedges, lies below you. Other islands still to be explored unfold in the sweeping panorama of scenic beauty: St Mary's, St Agnes and the Western Rocks, and there, seemingly just a stone's throw away lies Samson, its sandy shores highlighting the smooth rounded contours of its two low hills; here history and real peace awaits you.

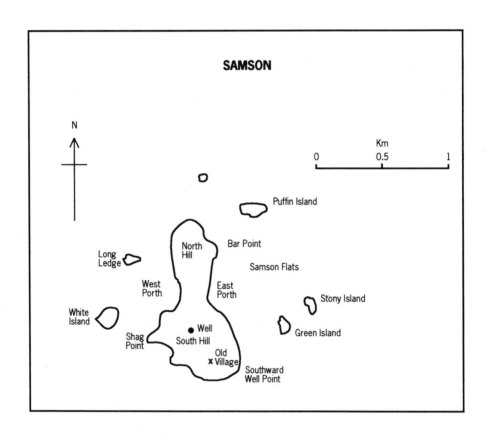

Plate 1   Sunset over Round Island, Page 44   Author

Plate 2   Oystercatchers on the dunes, Page 77   Painting by John Hamilton

Plate 3   Tean Sound, Page 43   Painting by Frank Wootton, OBE

Plate 4   Sea Thrift on White Island, Page 16   Author

Plate 5   Stonechat, Page 15   M. Hollings

Plate 6   Fulmar Petrel at nest, Page 14   Author

Plate 7   Summer on Tresco (Terns), Page 69   Painting by John Hamilton

Plate 8   Common Tern, Page 17   M. Hollings

Plate 9   Common Blue on finger, Page 24   Author

Plate 10   Belladonna Lily on St Martin's, Page 21   Author

Plate 11   Bermuda buttercup, Page 21   Author   Plate 12   English Catchfly, Page 22   Author

Plate 13   Pittosporum on St Martins, Page 20   Author

Plate 14 Yellow-Horned Poppy, Page 36 Author   Plate 15   Oystercatcher's nest, Page 45   Author

Plate 16   Gannet in flight, Page 31   M. Hollings

Plate 17   Small Tortoiseshell, Page 24   Author

Plate 18   Ruins of early settlement on Nornour, Page 29   Author

Plate 19   The old well on Tean, Page 44   Author

Plate 20   Tean Blue, Page 50   Author

Plate 21
Wild Gladioli
St Martin's
Page 21

Author

Plate 22
Lesser Black-Backed
Gull's nest
(Page 56)

Author

Plate 23   Clouded Yellow, Page 39   Author

Plate 24   Sea Rocket, Page 48   Author

Plate 25   Rock Spurrey, Page 49   Author

Plate 26   Lesser Black-Backed Gull at nest, Page 56   Author

Plate 27   Lesser Black-Backed Gull chick, Page 57   Author

Plate 28   Bluebells on St Martin's, Page 26   Author

Plate 29   English Stonecrop, Page 61   Author

Plate 30   Puffins socialising (breeding plumage), Page 58   M.Hollings

Plate 31   Razorbill, Page 59   M.Hollings

Plate 32   Common Scurvy-Grass, Page 62   Author

Plate 33   Pest House on St Helen's, Page 55   Author

Plate 34   Common Guillemot, Page 60   M. Hollings

Plate 35   Ringed Plover, Page 68   M. Hollings

Plate 36   Top of St Helen's looking to Tean and St Martin's, Page 64   Author

Plate 37   Hottentot Fig on St Helen's, looking towards Men-a-vaur, Page 62   Author

Plate 38   Tresco Abbey and Abbey Pool, Page 68   Author

Plate 39   Beach on Tresco from Lizard Point, Page 77   Author

Plate 40   Mute Swan at nest, Page 70   Author

Plate 41   Common Shelduck with young, Page 72   M. Hollings

Plate 42   Button Daisies or Chrysocoma on Tresco, Page 77   Author

Plate 43   Button Daisy or Chrysocoma, Page 77   Author

Plate 44   Burnet Moth on Lesser Hawkbit, Page 80   Author

Plate 45
Haresfoot Clover,
Page 76

Author

Plate 46
Sea Milkwort,
Page 78

Author

Plate 47   Bog Pimpernel, Page 78   Author

Plate 48   Purple Loosetrife. Page 78   Author   Plate 49   Yellow Bartsia. Page 79   Author

Plate 50   Spring Beauty, Page 21   Author

Plate 51   Spring Squill, Page 92   Author

Plate 52   Eyebright, Page 92   Author

Plate 53   Great Pool and Norrard Rocks, Bryher, Page 87   Author

Plate 54   View from old ruins on Samson, Page 99   Author

Plate 55   Gladden, Page 104   Author

Plate 56   Rosy Garlic, Page 22   Author

Plate 57   Shag, Page 111   M. Hollings

Plate 58   Tree Mallow on St Martin's, Page 115   Author

Plate 59   Old ruins and Foxgloves on Tean, Page 44   Author

Plate 60   Old Lighthouse on St Agnes, Page 120   Painting by John Hamilton

Plate 61   Small Copper, Page 128   Author

Plate 62   Common Mallow, Page 138   Author

Plate 63   Cretan Mallow, Page 138   Author

Plate 64   Wall Oxalis, Page 142   Author

Plate 65   Flowers of The Islands; Yellow Horned Poppy,
Page 36, Rock Samphire, Page 49, Sea Mayweed, Page 103,
Dwarf Pansy, Page 48, and Lesser Hawkbit, Page 76   Painting by Pat Donovan

Plate 66   Flowers of The Islands; Sea Holly, Page 48, Sea
Bindweed, Page 79, Sea Spurge, Page 48, Sea Rocket, Page 48,
Sea Storksbill, Page 62 and Sea Purslane, Page 126   Painting by Pat Donovan

8

SAMSON

Samson as seen from Tresco

To me Samson is the loveliest of the uninhabited islands, with its
sandy shores, seabirds and variety of colourful flowers. There is no
jetty on Samson, so your boat, the *Nimmo* from St Martin's, lands
you on the gently shelving beach of East Porth on the north-east
side of the island. The boatman agrees to pick you up in six hours'
time at the same place, when the tide will be halfway up the beach
and going out. It is early June, the sun shines fitfully from behind
high cirrus clouds and a stiff south-westerly breeze promises bad
weather to come; never mind, this is a small island of only some 95
acres and you have plenty of time to explore.

Samson consists of two hills, each just over 35 metres high,
joined by a sandy isthmus giving it the shape of a figure-eight. This
shape may have given the island its name, for the old Norse word
sammans-on means 'joined together' – a legacy from the past when
the Vikings roamed the seas around Britain in the eighth and ninth
centuries. Scillonians have a straightforward logic when giving
names to places and so Round Island and Black Rocks are very
descriptive; it is no surprise therefore that the two hills on Samson
are called North Hill and South Hill.

As you climb up the path leading to the top of North Hill your
mind wanders over the many aspects of the island's history;

clearly, two thousand years ago Samson was joined to Tresco and Bryher and at that time and for many years before it was well inhabited. Evidence of its earlier inhabitants is visible in the numerous entrance graves situated on both of Samson's hills. One particular example on North Hill is a substantial structure, revetted by kerbstones, within which is a well-preserved chamber roofed by two capstones. Excavations of this in 1930 revealed flints and Neolithic pottery but no human remains; the traces of ancient field systems can be found all around the island, particularly to the south-west. There may have been an early monastery here but the walls of more recent dwelling places are very evident around South Hill. Most of these were built in the seventeenth and eighteenth centuries. The first family settled here in about 1669, and by 1715 there were 12 inhabitants; the numbers gradually increased until in 1822 there were seven families with a total population of 34. These hardy people survived by fishing, kelp-making, pilotage and growing corn on the 'neck' between the two hills where under the sandy topsoil there is a fertile clay. Water was always a problem even though there were two wells on the island, one near the base of the northern side of South Hill, and the other called Southward Well, just above the shore in the south-east; but quite often these produced little or no water at all during a prolonged summer drought.

The demise of the islanders was hastened when a tragedy occurred during the Napoleonic Wars of 1803–15. A small French barque was seen becalmed and drifting just off Samson and the islanders thought they would benefit by boarding her and handing her over to the authorities on St Mary's. They rowed out in three gigs, overcame her crew of ten and reported the capture to the garrison commander; they were then advised to sail her to Devonport, where they would be better rewarded. So they set off for the mainland, only to founder on the Wolf Rock just off Land's End, which then was not marked by a light. Sadly all ten Frenchmen and nineteen Samson men were drowned, an unwelcome blow for the people of the island. Many dispute the truth of this story but there is some evidence that it is not entirely legend. In any event, by 1850 the population of Samson was down to three or four households and their life was miserable in the extreme, all of them finding it impossible to live off the land. Augustus Smith, seeing the hopelessness of this island community fighting a losing battle for survival, decided that they should be transported across the water to St Mary's. So in 1855 they left their homes, and the north slope of South Hill was enclosed by a boulder wall and used as a deer park; but, alas, this experiment of innovative twentieth-century farming failed when the deer escaped across Samson Flats at low tide to Tresco.

So the island was once again left to the seabirds and the rabbits, who bred undisturbed on the hillsides. Bracken and bramble quickly took over, unchecked by humans or deer, and now these predominate and make walking off the tracks hard going; they surround the old cottage ruins and climb up and over the ancient walls, making them almost invisible in places. Limpet shells scattered in piles here and there beside the ruins indicate sites of 'kitchen middens' or waste tips where the inhabitants discarded the remains of one of their essential sources of food. What stories these ruins could tell of hardship and poverty; yet at the same time there must have been great pleasures when fishing was good, food was in plentiful supply and the setting sun over the wide horizon of blue sea gave an atmosphere and peace that sustained them when the times were tough. Time and the measured rhythm of the seasons must have been all-important to these people, for time marks the birth of life itself, and death when it departs. As they lived they would have found happiness and unhappiness, ecstasy and torment, excitement and boredom; all feelings that must inevitably be found together, for one cannot exist without the other and each is dependent one on the other. Worth thinking about as you sit alone on top of Samson's hills and contemplate the history and atmosphere all about you.

Feel the warm breeze on your face as you look out to the darkening sky in the west as black ragged clouds of nimbostratus lower to the horizon over the Western Rocks. Glance down at some low-lying rocks just offshore and with your binoculars begin to count the *common terns* nesting on the rocks. Over 100 pairs breed in various scattered colonies on the islands but the majority are to be found here on the grass-topped rocks of Green Island, sheltered somewhat from westerly gales by the bulk of South Hill. See if you can spot one of only about six pairs of *roseate terns* or perhaps a *Sandwich tern*, which has only recently started to breed here on Scilly.

The roseate tern can be distinguished from other terns by its particularly long tail streamers, and in the breeding season by the lovely rosy flush on its breast. It differs from the common tern by having red rather than black legs, and by having a largely black bill, whereas that of its relative is dull red with a black tip. Overall it has a whiter appearance than the other species. It nests in complete harmony amongst the common tern colony, its two or three eggs being laid on the ground in a slight hollow; these eggs are indistinguishable from those of the common tern, described in Chapter 2.

The sandwich tern is easily spotted by the yellow tip to its black bill, and when on the ground it displays elongated crest feathers on the back of its crown, giving it a shaggy appearance. Its eggs, like the common and roseate terns', are laid in May and June on the bare ground and are only distinguishable by their slightly larger size, about 53 millimetres long compared to the 44 millimetres of the other two species.

Like many sensitive areas on Scilly, it is forbidden to visit Green Island from 15 April to 20 August without a special permit but it is quite possible to view them from Samson itself. Look there, as all the terns rise in perfect unison to ward off a *greater black-backed gull* that has got too close to their territory. They whirl around like a shower of snow gently blown about by fickle gusts of wind, all the time uttering harsh grating calls, 'aaak-aaak-aaak', and angry, chattering 'kikkikkikik'. As the gull is chased away the terns quickly settle back down to their nests and calm descends once again on this peaceful scene.

As I write, *lesser black-backed gulls* have taken a great hold on the southern slopes of Samson. Their nests are scattered everywhere in colonies of a dozen or so and you walk amongst them at your peril, the large white birds screaming and diving at you from all directions. Occasionally a tap on your head with their feet hastens your departure. I wonder if all the brown rats have been exterminated on this island too, just as has happened on St Helen's; it would be as well that they had, as otherwise they would pose a distinct threat to the terns on Green Island since the short stretch of water at low tide would be of little hindrance to the swimming proficiency of this vermin.

As you make your way across the island to its south-west tip at Shag Point you hear the by now familiar plaintive cry 'kitti-wee kitti-wee'. A colony of *kittiwakes* have their nests on the low cliffs, and as adult birds fly in from the open sea, their pure white wings with black tips are highlighted against the threatening dark clouds now fast approaching.

You seek shelter beside an old cottage wall just as a *blackbird* puts out its splendid flute-like song from the top of a nearby bramble bush. Notice that its bill is a distinct red, quite different from that of its mainland counterpart, which is yellow. It is thought that the Scillonian blackbird has developed this characteristic feature by eating the berries of the *pittosporum* hedges which contain a red dye. This particular bird could just be passing through the islands but more likely he has a mate and a nest in the area. The blackbird is one of the earliest birds to breed, building its nest of grass and mud about 1 or 2 metres above the ground in a low growing bramble bush or on a ledge in an old wall. Unlike the song thrush described in Chapter 6, its nest is lined with grass and its four to six eggs are a dull bluish-green, spotted and blotched with reddish-brown and grey. The first eggs are laid in February and after several broods the last ones can still be found as late as August.

In the early spring the old field walls can be easily seen before the bracken and brambles grow up to hide them. Within the enclosures lie the fields which the islanders cultivated for hundreds of years, and the soil here, being deep and fertile, supports a luxuriant vegetation. *Bluebells* predominate in these ancient field systems and seem to avoid places where the soil has not been cultivated. These plants, which are abundant on all the islands, are known as *cuckoos*, possibly an ancient name indicating the arrival of that bird when the flower just begins to bloom. Bluebells store up a supply of food in the yellowish-white bulbs, which remain dormant during the summer and winter, buried deeply in the earth. Early in March long, narrow, deeply channelled leaves appear and push through, sometimes piercing old dead leaves, grass and bracken. Soon these first bluebell leaves lengthen and turn a richer green, and at their centre what was once an empty tube fills up with a flower spike of closely pressed buds; rapidly the whole area of ground is filled with bluebell spikes all pointed upwards amongst a sea of dark green leaves. What a sight this splendid flower of the spring makes when viewed in Samson's setting of blue sea and sky and white sand.

In sheltered places among the bracken and the bare rocks around the ruins of South Hill, the pale yellow flowers of the *primrose* spring up from their loose rosette of crinkled leaves. On Samson this early spring flower is thought to be a native plant and Scillonians once used to gather the roots and flowers and transplant them to other islands. Apart from here on Samson, the primrose is a very local plant in Scilly and is only found on the dunes near Carn Near on Tresco, and in the Old Town churchyard on St Mary's, both places where they have been planted. Like the bluebell, the primrose stores up sufficient foodstuffs, chiefly sugar and starch, in its thick fleshy roots and as soon as the winter is over it is ready to display its attractive flowers.

June is the best time of year to find and admire the many different plants on Samson, for of all the uninhabited islands on Scilly, this one has the greatest variety, with about 125 different species. Descending the path of South Hill, you suddenly come upon the well where water trickles out over a rock and into a small basin; here the water, about 30 centimetres deep, is crystal-clear and is contained by a semicircle of boulders. Look into the black pool and see your reflection etched neatly on its surface; the rosette of elliptical floating green leaves of the *water starwort*, an occasional plant in Scilly, may attract your attention. Pick one gently from the surface and see the tiny green petalless flowers at the base of the leaves; these are in bloom from April to September. Growing nearby in the mud at the pool's edge is another tiny flower, *blinks*; this inconspicuous plant has long straggling stems on which pairs of oval leaves grow on opposite sides. The tiny five-petalled white flowers bloom from March to October and are held in loose clusters at the end of long stalks. Groups of ferns also hang their green fronds over the granite rocks and the fleshy round leaves of *wall pennywort* can be seen growing out of a crevice between the boulders. What an enchanting place this would be on a hot sunny day, a place where the shade of the steep rocky hillside would keep you cool and refreshed. How many times did people of this island quench their thirst and sit here for a while to rest and contemplate the mysteries of life itself? Truly this is a small Garden of Eden set amongst the superb splendour of the island. Time to move on as a gentle mist just touches the top of the hills with a soft caress and Tresco, so near across the shallow water, starts to disappear in low cloud. A slight twinge of apprehension fills your mind; will your boatman be able to extract you from the island? You feel the hardness of the chocolate bars in your pocket, the weight of food and thermos in your rucksack, and then remember the reassuring words: 'Don't worry, I shall be back at five o'clock to pick you up, whatever the weather.'

Look at those tall evergreen bushes with greyish feathery foliage growing around the ruins. This is *tamarisk*, a native of the western Mediterranean, which was one of the first to be planted as a windbreak in Scilly; it gives good resistance to cutting salt-laden winds and was used quite extensively on Samson around the settlement. It grows as high as 4 metres and has reddish stems and minute evergreen leaves. The pale pink flowers are minute, only 2 centimetres in diameter, and bloom from June to as late as November. The stems of tamarisk are very pliant and are used in the making of lobster pots.

As you walk along the beach on the south-eastern shoreline,

look out for a rare plant, the *greater skullcap*. This grows along the shore by Southward Well pool, its only site in Scilly, where it was first recorded in 1864. About 15 to 50 centimetres tall, the greater skullcap has bright blue flowers less than 2 centimetres long, which appear in pairs from the stem at the leaf base. The flower sheath resembles a Roman soldier's leather skull helmet, hence its name. The leaves are lance-shaped and have smoothly rounded teeth. You will find this plant in flower from June to September.

Nearby you will find the spikes of inconspicuous green, petalless flowers of the *red goosefoot*. This somewhat local plant is found on only a few of the islands and can be recognised by its reddish stems and shining, fleshy, triangular-shaped leaves. The whole plant grows to some 50 centimetres in height and should not be confused with two other more common allied species: *fat-hen*, with greyish green foliage, and *nettle-leaved goosefoot*, which has more deeply toothed leaves of a brighter green. Both these latter species grow in cultivated land and waste places on the inhabited islands but are not recorded as growing on Samson.

On the sandy wasteland between the two hills, look for the *small bugloss*, a somewhat untidy, bristly plant up to 30 centimetres tall. It is a common weed of the inhabited islands and here on Samson it has its only location on the uninhabited islands. Its small bright blue funnel-shaped flowers bloom from May to July; look carefully at the corolla tube beneath each flower and see its characteristic double bend. Another plant characteristic of this sandy area is the yellow-flowered *biting stonecrop*, mentioned earlier in Chapter 5. Look for it as it hugs the ground with its thick, fleshy leaves; it is easily recognised by the bitter, peppery taste when a small piece is cautiously chewed.

One of the commonest and most conspicuous plants of Samson's foreshore and wastelands is the *scentless* or *sea mayweed*. The large daisy-like flowers, some 5 centimetres in diameter, appear in May and last until late autumn. Each flower head is supported by a long smooth stem which is clothed with alternate stalkless leaves, each finely cut into many slender segments. As its name suggests, the flowers have no scent except a slight pungent odour; this distinguishes it from *chamomile*, which is common on the inhabited islands and has a distinct sweet and aromatic smell. The sea mayweed is one of the few plants that is found growing amongst the stones on the shore, a place where little else can get a hold. How pleasing it looks in this setting of grey and pale-pink rounded rocks and brown scattered seaweed.

As you take the path leading to East Porth, notice many tall plants of the *balm-leaved figwort*, with its distinctive heart-shaped, wrinkled leaves, which you remember seeing on St Helen's; but what is this iris growing in clumps amongst the grass beside the track? This is called *gladden* and is an uncommon plant, found elsewhere in Scilly only on St Mary's, Tresco and St Agnes, where you will see it on the sand dunes, rough cliff slopes and lanesides. It has distinctive, long, dark green lance-shaped leaves, which quickly turn brown and yellow at the tips, scorched by the salt-laden winds. Its somewhat modest pale purple flowers are 6 centimetres across and have distinctive dark veins running through them like a cobweb. They bloom from May to July, and the seeds when ripened are a brilliant orange and lie in rows like peas in a pod on the open segments of the seed capsule. Crushed between the fingers, the gladden gives off an unpleasant odour of stale beef, hence its other names, *stinking iris* and *roast-beef plant*.

A drizzle has set in and you notice how soft and warm it seems as you feel your hair gradually getting soaked and small rivulets descend down the back of your neck. You put the hood of your anorak over your head, knowing it will restrict your vision and impair your hearing. It is no good looking for butterflies; they will be well hidden and lying low, deep amongst the grasses and bracken. But here on the path is a slow-moving black beetle with long slender spatula-like legs. This is one of the 700 or so species of *ground beetle* to be found in the British Isles. Although they have rudimentary wings, most are fused together and they cannot fly. With powerful jaws, these ground beetles are ravenous predators and feed on small worms and caterpillars, usually coming out at night to feed. During the day they remain hidden under stones or in thick vegetation; the beetle's life history goes through all the stages of egg, larva, pupa and adult insect.

On the pink *thrift* flower you will see another beetle, this time very attractive, with its metallic green body flecked with white; this is the *rose chafer beetle*, whose fat white grubs live off decaying wood or vegetation. These larvae are long-lived and take about two or three years to reach the adult beetle state after pupating. Unlike the ground beetle, the rose chafer can fly, using its two hind wings; the forewings are hardened and give protection to the body and cover the delicate hind wings when they are not in use. Although the rose chafer commonly feeds on rose petals, it also eats the flowers of the thrift and *Hottentot fig*. In Scilly there is a black variety of this beetle which is only found elsewhere in the Channel Islands and Corsica.

You hear the noise of the *Nimmo* through the mist and soon its small dinghy is nosing in through the gentle waves to pick you up; a black and white mongrel sitting on the prow greets you with a friendly bark and a wagging tail. As the boat leaves, you know that beneath you are the remains of boulder walls of an ancient field system that is uncovered at low tide and can be seen through the clear water on a fine day from on top of Samson's hills. But today the scene is bleak and cheerless, no deep blue sea and sky that one expects in summer on Scilly, just a gentle soft drizzle and a mist that now obscures the island itself. You recall the thoughts of ancient Greek writers you were made to study many years ago; pleasures are transient, and like happiness you must make the most of them when they are encountered and learn not to expect them all of the time. Samson in this weather is certainly not as pleasurable as it could be, but nevertheless it has a desert island quality, where time stands still and where you can experience that rare chance of undisturbed peace, and take in its atmosphere and its sense of history. As the drizzle gives way to heavy rain and your boat ploughs its way through the grey water, you look forward to the pleasures of the hotel and a warm bath; time to contemplate your next journey out to the Western Rocks, lying as stranded remnant hilltops of a bygone age. The names of these rocks conjure up so much of the savagery of the surrounding seas: Hellweathers, Crim, Crebbinicks, Crebawethan and Minmanneth; that Scillonian logic once again no doubt! Enigmatic and somewhat incongruous are the names of others in the area: Daisy, Jacky's Rock, Jolly Rock, Rags, Silver Carn and Brothers, to name but a few. What stories do these names conjure up of the happenings around these remoter Isles of Scilly?

Kittiwake

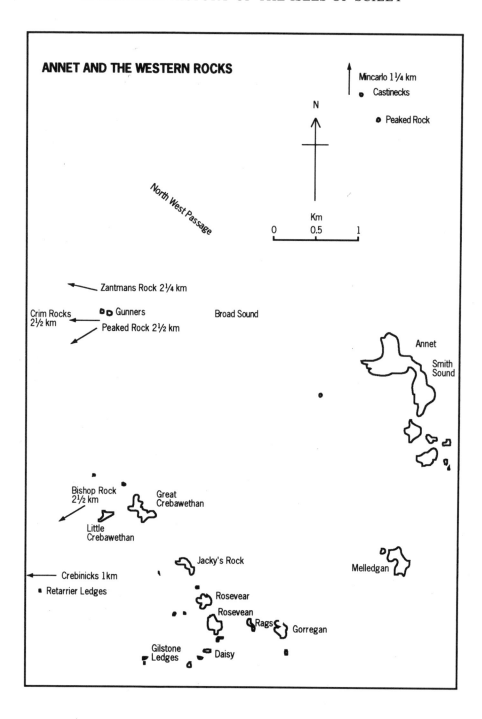

ANNET AND THE WESTERN ROCKS

Mincarlo 1¼ km

Castinecks

Peaked Rock

N

Km

0   0.5   1

North West Passage

Zantmans Rock 2¼ km

Crim Rocks
2½ km

Gunners

Peaked Rock 2½ km

Broad Sound

Annet

Smith
Sound

Bishop Rock
2½ km

Great
Crebawethan

Little
Crebawethan

Jacky's Rock

Melledgan

Crebinicks 1 km

Retarrier Ledges

Rosevear

Rosevean

Rags

Gorregan

Gilstone
Ledges

Daisy

9

ANNET AND THE WESTERN ROCKS

Workman's Hut, Rosevear

The sun shines out of a clear blue sky and as you look out from the hotel you see that the white water out at Black Rock and Round Island is less threatening. Clearly the wind of the last 24 hours has moderated and it seems a reasonable day to venture out to the Bishop Rock Lighthouse and the Western Islands; it may be possible to land on Annet too.

Past Kettle Point and through the narrow channel between Tresco and Bryher and then out into the vastness of the North West Passage; despite the light breeze, the Atlantic swell is there as ever, the boat rising and falling with the rhythm of the waves, occasionally broken when the bow hits hard into one out of sequence. The lighthouse on Bishop Rock stands out as a lonely sentinel against the distant horizon; Mincarlo, black and uninviting, is to starboard and friendly Samson with her two green hills falls away to port. Soon the stark sinister pinnacles of the Crim Rocks can be seen doused in spray.

As you look out to the horizon, never near but always far, you may remember the stories of the many ships that came to grief on the many barely submerged reefs that abound in the area, and the words of William Shakespeare seem apt:

Methought I saw a thousand fearful wrecks;
A thousand men that fishes gnaw'd about;
Wedges of gold, great anchors, heaps of pearl,
Inestimable stones, unvalued jewels,
All scatter'd in the bottom of the sea.

(Clarence's dream, *Richard III*)

You may recall the story of the greatest peacetime disaster in the Royal Navy's history when, on the night of 22 October 1707, four large ships of the fleet of Rear Admiral Sir Cloudesley Shovell struck the Western Rocks and some 1,670 men were drowned. The Admiral, a national hero of his day, was aboard his flagship, the 90-gun, 1,459-ton *Association*, returning home to Portsmouth from the Mediterranean with gold and silver coins, silver plate, bronze cannons and the like. Due to an extraordinary error of navigation the *Association* went aground on the Gilstone Ledges and sank. According to some accounts, Sir Cloudesley managed to get away in a ship's boat with his pet greyhound and some treasure but then came to grief just south of St Mary's, and his body was found washed ashore at Porth Hellick. It was interred in a shallow grave on the sandy foreshore but later removed, and after a state funeral given by Queen Anne he was laid to rest in Westminster Abbey. A stone monument and plaque now marks his original burial place at Porth Hellick.

Of the other three ships it is thought that the 70-gun *Eagle* went down amongst the Crim Rocks and the 48-gun *Romney* sank on the Crebinicks. The 8-gun fireship *Firebrand* was holed and eventually sank in Broad Sound. Although much salvage was done at the time, much to the advantage of many Scillonians, it was not until 1967 that divers rediscovered the wreck of the *Association* and recovered quantities of treasure, some of which can be seen in the St Mary's museum.

Almost on a par with the disaster of Sir Cloudesley Shovell and his fleet was the sinking of the *Schiller* in 1875. She was an illustrious German steamship, one of the largest of her day, and on passage from New York to Plymouth with 377 passengers and crew. In dense fog the ship struck the Retarrier Ledges just to the south-east of the Bishop Rock Lighthouse, and no less than 311 passengers and crew lost their lives. At the time she struck the rocks a social function was in progress and most of the passengers were dressed in their finery; panic ensued as everyone rushed for the lifeboats. The captain tried to restore order by firing a pistol above the heads of the frantic crowd but to little effect. The funnel came down, crushing one of the boats, and of the eight available only three were launched. One can hardly imagine a more terrifying scene. Many bodies were washed ashore on the islands and a sad procession of 20 carts filled Hugh Street for the funeral at the Old Town churchyard on St Mary's.

Of course there were many, many other ships that came to grief in the area of the Western Rocks. Notable was the 700-ton Dutch East Indiaman the *Hollandia*, which came aground on the Gunners Rock and sank just north of Annet in 1743 with the loss of all hands. It was on its maiden voyage from Holland to the Far East, carrying, amongst other valuables, a cargo of Spanish-American silver coins minted in Mexico City by a special screen press installed by the Spaniards in 1732. This press produced well-refined coins with a very pure silver content and fine milled edging, and these were known throughout the world as 'pieces of eight'.

The advent of modern sub-aqua diving has led to the discovery of the remains of many of these ancient wrecks and the recovery of many valuable artefacts, jewellery and coins – again Shakespeare's words are so masterful and adroit in their description.

Suddenly your attention is drawn to a large flock of *gannets* just a few hundred metres off the port bow. What a sight they make against the blue sky; like great white paper darts, they plunge into the heaving blue ocean, not in deliberate unison but each bird one after the other in a random fashion. Splashes of white water appear on the surface where they hit and swim down to chase fish beneath the waves.

A pair of *cormorants* take off in rapid flight from the water just ahead of the boat, their necks extended just slightly above the horizontal. The Bishop Rock Lighthouse looms up large and straight, a lone sentinel standing on the most distant and outermost pinnacle of rock in this group of islands. Two sites were short-listed for the construction of this much needed light to mark these dreaded Western Rocks; the Bishop Rock itself and Rosevear, a small rocky islet just over 3 kilometres to the east. The Bishop Rock was chosen but Rosevear was used as a base for the men engaged in constructing the lighthouse. So from 1840 to 1857 a gang of Cornish workmen lived on Rosevear almost continuously in purpose-built stone huts. These hardy workers and craftsmen put the finishing touches to huge granite blocks so that they would fit onto the wave-swept rock, fastened together with iron pegs. The original blocks were hewn out of Cornish quarries and then roughly fashioned on the quayside on St Mary's before being transported to Rosevear. The first structure built on the rock was made of iron, open at the base to allow the waves to pass through, but this was washed away in a storm on 5 February 1850; luckily it was unmanned at the time and no lives were lost. It was then decided that the whole lighthouse should be built of stone. As they worked, often the men securing the lowest stone blocks were swept off the rock, but a waiting boat always rescued them. Remarkably, no one died or was seriously injured during its construction despite the hazardous nature of the work. The lighthouse was first lit on 1 September 1858, its light of two white flashes every 15 seconds can be seen in clear weather at a distance of 42 miles; a foghorn blasts out whenever the visibility is poor. There is no doubt that the light-house did much to prevent shipwrecks, although it is ironic that in 1901 the 2,867-ton four-masted *Falkland* bound for Falmouth actually came aground on the lighthouse itself.

The completion of this magnificent lighthouse, perhaps the most exposed in the world, was described by Prince Albert as a 'triumph of engineering skill and perseverance'.

At times the storms were so great that waves washed right over the top of the structure and it swayed so much in the wind that crockery fell off the shelves. It was not surprising, therefore, that in 1882 a thick cylindrical base was built around the original founda-tion and its height was increased by 11 metres to its present height of 53 metres.

A tribute must be paid to those men who constructed the great Bishop Rock Lighthouse. What hardships they must have endured on their tiny isle of Rosevear; they had their own vegetable garden and at times, when they were cut off from St Mary's, they were reduced to eating limpets. They apparently had some social life as dances were held on the island on the first Saturday of each month when the ladies came out from St Mary's and St Agnes, weather permitting! Certainly it is known that the occasional ball was held by Rosevear's robust and intrepid inhabitants, when their living quarters and work sheds were decorated and they danced with their invited guests to the sound of music mingling with the soft wailing of the wind and the booming of the waves on the rocks. These men must have enjoyed one of the most romantic of all settings as well as having endured great hardships.

As the boat swings around the great base of the lighthouse, her beauty now spoilt somewhat by the helicopter platform built in 1976 around her top, you can see the cracks down the side between the granite stones. These are wider near the top and have been caused by the great storms of recent times. Even her huge solid bronze doors had to be replaced in 1997 after they had been smashed in by huge waves.

The boat now heads east towards the Western Rocks, those islets of grey granite touched here and there with patches of green where *sea beet* and *common scurvy grass* have secured a hold. Here these bastions of rock set amongst the deep blue sea have remained a guardian to the approaches to Scilly for millenniums past, with nothing but the ocean between them and the Americas some 5,000 kilometres to the west. The sea sucks and seethes around the black base of these reefs; it is relatively calm now but one still feels the awesome power of the ocean and it is very easy to imagine the scene during a storm when the entire area is white with foam as great breakers crash down on the rocks with a deafening roar like thunder. As the boat cautiously approaches the islands, *grey seals* appear almost by magic, their heads showing for a moment on the crest of the swell and then disappearing again. *Razorbills* and *guillemots* bob up and down with the waves, looking completely at ease with the hostile sea.

Unexpectedly, a metre or so down in the clear green water are large numbers of opaque polythene carrier bags, their seemingly unused appearance and mint condition giving a sense of unreality to this maritime environment. You may remember the news of 26 March 1997 of the wrecking of the German-owned ship *Cita* on rocks off St Mary's. Much of its cargo of toys, plastic film, carrier bags and the like were washed out to sea, to create a real hazard to marine life. We live in an age where greed and the pursuit of material riches seem to be the aims of so many people who in their selfishness neglect the environment and sow the seeds of man's ultimate destruction; for the resources of our fragile planet are finite and surely cannot go on for long meeting the needs of an increasing number of people eager for more and more. Here in these beautiful islands of Scilly the pressures for development and change grow stronger, and thank goodness The Isles of Scilly Environmental Trust has been set up to manage and educate people of the need to protect the uniqueness of the Scillonian heritage. Let us hope that not too many of the seabirds we see before us now get caught up and choke on the polythene detritus just below the surface.

We pass by Jacky's Rock and soon the gaunt gable ends of the workmen's houses on Rosevear come into view. Over 120 years ago this was a place of hard toil for those men who built the Bishop Rock Lighthouse. A group of *shags* stand in a line on a lichen-covered outcrop facing the sea, their wings held wide for the breeze to blow through their feathers to dry them. There are about 1,100 pairs of shags in Scilly, a number that greatly exceeds the 50 nesting pairs or so of their larger relative, the *cormorant.*

The shag is about 75 centimetres long and has dark bronze-green plumage without white markings and has a distinct crest of feathers on its head. The cormorant is larger, about 90 centimetres long, has darker plumage and can easily be distinguished from a distance by its noticeable white chin and cheeks and by its thicker and heavier-looking bill. Both species are marvellous swimmers and divers, pursuing fish deep under water and catching them with ease. The shag prefers to hunt for small rockfish of little commercial value off rocky shores and in deep water; whereas the cormorant is not popular with local fishermen because it prefers flatfish, which it looks for on the sandy bottom of the shallower waters. Interestingly, the Japanese and Chinese have trained cormorants to catch fish for them. The birds are taken out to the fishing grounds in small boats; rings are slipped around their necks to prevent them swallowing their prey and they are sent overboard. Unable to swallow the fish they catch, they swiftly return on board to disgorge them before the delighted fishermen. They then return quickly into the water to seek more fish.

Both shags and cormorants nest on rocky islets, the more common shags tending to build their nests in colonies in crevices and beneath overhanging rocks and sometimes on ledges or amongst boulders on rock-strewn beaches. The cormorant too builds its nest in small colonies but in more exposed positions, on cliff ledges and on top of large rocks. The nests of both birds are similar, being made of sticks (particularly of tree mallow stems), coarse grass and seaweed, and in March or April both species lay from two to five pale green eggs. These are covered with a thick coating of chalk, which soon becomes discoloured from the droppings of the birds, and the nests start to smell foully of decaying fish. Only the size of the eggs distinguishes one species from the other, those of the cormorant being the larger, of course. The eggs hatch out in about 24 days in the order that they were laid; the young are therefore of different sizes and it is quite usual for the eldest to claim the main share of the food which is regurgitated by their parents. What a sight to see the young almost disappearing down the brilliant yellow mouth of their parents as they hungrily search for their food. When food is scarce, the older and stronger fledglings survive, leaving the younger to perish; nature's way of ensuring that at least some of the brood achieve adulthood to perpetuate the species.

Young cormorants and shags mature rapidly and are able to fly well at six weeks old. The ability to fly well early in life is seemingly important to their survival, for the plumage of these juvenile birds is not yet waterproof and if caught out in the deeper waters by rough weather, they easily escape drowning by flying to the shelter of rocks.

The boat heads north towards Annet, past Gorregan, the only other nesting site for the *guillemot*, apart from Men-a-vaur, and past Melledgan, one of the main nesting colonies of the cormorant. As you near Annet you can see the vast numbers of birds in flight above it, for this is the principal bird sanctuary of Scilly and permission from The Environmental Trust is necessary to visit the island during the breeding season. Annet is a low-lying island of just over 50 acres, rising to only 18 metres on its northern tip; its coastline is deeply indented and most of its beaches are strewn with granite boulders of all sizes. The island was frequented by prehistoric man and his presence is evidenced by the remains of hut circles and field systems. Certainly, man could not exist on the island today, for now with much higher sea levels great breakers wash right over it in storms. In the early nineteenth century sheep and cattle were grazed here and bracken was cut for fuel by the people of nearby St Agnes.

For the ornithologist the main attractions on Annet are the *storm petrel* and the *Manx shearwater*. There are several hundred pairs of storm petrels nesting in colonies in rock cavities on certain of the beaches, but as we noted in the full description of the bird in Chapter 4, it is very seldom seen, flitting into its colony at dusk and out again at dawn.

The Manx shearwater, so named because it used to breed in large numbers on the Isle of Man, is another seabird species that is rarely seen at its breeding colonies as it is strictly nocturnal, taking care to come ashore only under the cover of night. This bird spends most of its life ranging across the Atlantic Ocean, regularly wintering off the coast of Argentina and returning to breed from May to October on remote islands in the north and west of the British Isles.

About 35 centimetres long, the Manx shearwater can be distinguished by its sharply contrasting black upper parts and pure white underparts. It has a slender bill and can usually be seen in scattered groups way out at sea, gliding for long periods and veering from side to side, showing alternately black and then white. What a sight they make as they skim the wave tops on stiffly held wings and at times fly right through the white crests.

Unlike storm petrels, which also spend most of their life out in the deep ocean, Manx shearwaters do not follow ships. They congregate in large numbers just out to sea off Annet just before sunset in preparation for their flight to their nest sites. The shearwater is fully adapted to oceanic life and its feet are set well back on its body, a design giving it maximum propulsion in the water. However, this arrangement makes the bird ungainly on land and vulnerable to predators such as black-backed gulls, and it is for this reason that its visits to land are at night.

The Manx shearwater builds its nest of dead bracken, bluebell bulbs, feathers and grass at the end of a 1- to 3-metre-long burrow which the bird has excavated under a rock or boulder. Parts of Annet are honeycombed with these holes. The single pure white smooth egg is laid in April or May and is incubated by both parents, each taking shifts lasting several days, but sometimes up to ten days at a time. During this time the 'off-duty' bird spends its time out to sea, feeding itself in preparation for its long fast on the nest. Ringing of breeding Manx shearwaters has shown that it travels as far south as the coast of Spain some 960 kilometres away to feed on sardines, a round journey of over 1,900 kilometres that would take them away from their breeding colonies for two days or more. After about 51 days the egg hatches and the chick is fed by the parents with regurgitated fish and plankton. After about 62 days the chick is abandoned by its parents and its body fat is greatly reduced as its feathers are hardened and set firm, and as it exercises them coming out of its burrow at night to stretch and flap them up and down. Eventually the abandoned fledgling realises that it must fend for itself or die of starvation; so after about 10 days on its own, when night falls, it sets forth on a hazardous journey, scrambling through bracken and over rocks to the sea. How does it find its way? Any mistake or any delay on its journey that results in it still being on land in daylight leads to certain death by marauding gulls. It is not surprising that at the end of the breeding season on Annet many well-picked carcases of Manx shearwaters can be found littering the foreshore. What happens to the young shearwaters after they reach the sea and are able to fly and dive for fish is a mystery. How long do they remain solitary at sea roaming the wide Atlantic? When do they mate and return to their breeding colony? Many questions are yet to be answered.

One of the real experiences is to spend an evening on Annet and wait until after the sun sets. A still, moonless night sees the shearwaters at their busiest, but they are in no hurry to return to their nests; they have spent the day off the island, resting and preening themselves on the water. It is after midnight when dark shapes begin to hurl themselves out of the sky and onto the ground. There is a chorus of groans and wild crowing from overhead, and from the burrows underground comes a low, guttural crooning. With birds flying in from the sea and others leaving their nests and scuttling about looking for a good take-off place, everything is hustle and bustle, and all around a noise like countless chickens and whimpering puppies.

But Annet is not just an island of Manx shearwaters; there are numerous pairs of *puffins*, *razorbills*, *shags* and in recent years an increasing number of *fulmars*. In addition there are some 600 pairs of *lesser black-backed gulls*, 100 pairs of *greater black-backed gulls* and 50 pairs of *herring gulls*, all of which create a hazard to the others. I have watched a greater black-backed gull, its huge black wings silhouetted against a grey sky, swoop down on a flight of puffins coming in with fish for their young, grab one in its mouth and then fly off, only a few feathers blown in the wind marking the point of impact. A few *wrens* and *blackbirds* nest on the island, as well as *oystercatchers* and one or two pairs of *ringed plovers*. No wonder this is a bird sanctuary whose existence must be protected from too much disturbance.

Over 50 species of plants are recorded from Annet, not a bad record for a small island exposed to the ravages of the gales sweeping in from the vast ocean. *Sea thrift* is predominant, forming huge hummocks particularly on the northern part of the island, and in May its freshly opened pink flowers present the visitor with a resplendent and colourful sight. Along the boulder beaches, the rare *shore dock* can be found, and the fleshy plants of *sea beet* and *common scurvy grass* are common; but tallest of all of them is the *tree mallow*, which grows in rows like hedges.

Tree mallow is unmistakable, with its tall, stout, tough and branching stems some 1 to 2 metres high; the stems are covered with soft hairs, as are the triangular lobe-shaped leaves. The large pinkish-purple flowers are held on stalks at intervals up the stems and have dark purple veins which widen out of the base of the petals to form a dark centre. The flowers bloom from April to September. Interestingly, as I have already mentioned, the stems of this plant are much used by cormorants and shags as material for building their nests.

In the spring, plants such as the *lesser celandine* and *bluebell* are abundant. The bluebell was covered in the chapter on Samson but this is the first mention of the lesser celandine, which is to be found everywhere in Scilly, under bracken, on cliff slopes and beside hedges. This well-known plant hardly needs much description, with its bright golden star-shaped flowers giving colour to the islands in the very early part of the year. Its heart-shaped leaves die down early in Scilly, so its presence is easily overlooked as the grass and bracken begin to grow. The lesser celandine has a fibrous rootstock which produces a large number of cylindrical tubers which readily break free and produce new plants the following year. These tubers were said to resemble haemorrhoids and in the seventeenth and eighteenth centuries herbalists believed in the 'Doctrine of Signatures', which proclaimed that God had given to every plant a physical clue as to its medicinal qualities; so the lesser celandine was used then to treat piles and it was known as *pilewort*. Interestingly, lesser celandine is common in churchyards and this gave it extra credence because herbal plants that grew in sanctified ground were considered to be especially powerful.

Annet is a fascinating island, with its quality of loneliness and peace, and yet you depart with added awareness of the dichotomy of nature, its great harmony and deep discord; and of the birds who are savagely aware of each other in their great fight to stay alive and to perpetuate their own. The boat takes you across the short stretch of sea called Smith Sound to another inhabited island, St Agnes. As you leave behind the Western Rocks and their colonies of seals and birds, the gently movement of the boat on the ocean's swell and the setting sun in its blaze of gold and red, all give an intense feeling of an undisturbed calm that remind me of the music of Mendelssohn, and particularly of the restful, singing romance of the second movement of his first piano concerto.

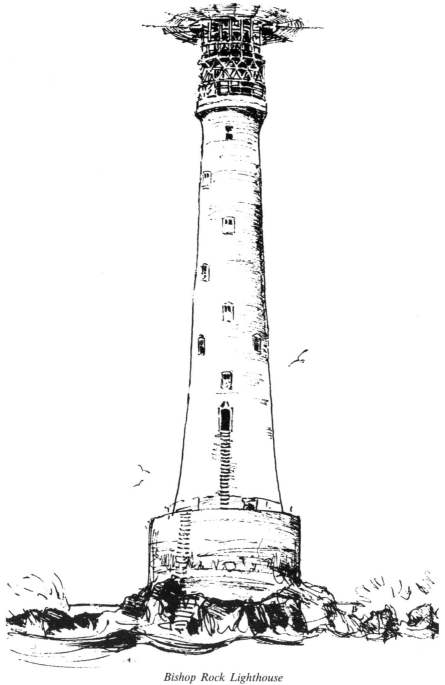

Bishop Rock Lighthouse

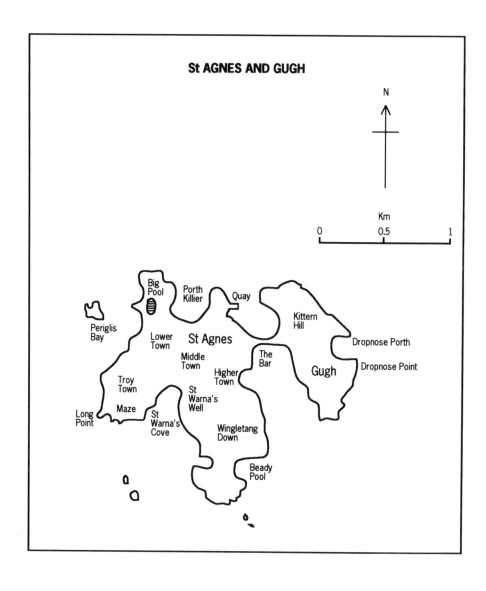

10

ST AGNES AND GUGH

The Lighthouse, St Agnes

St Agnes is surrounded by deep water, which in a sense gives it a different complexion from the other inhabited islands. A feeling of isolation and independence, accentuated by the remoteness of the rocky reefs off its western shores, is all pervading when one walks over its weathered granite headlands and ridges. Physically it is not dissimilar to the Western Rocks, being small, round and relatively flat, with a rocky coastline. The central granite ridge predominates and rises in places to just over 25 metres. The smaller island of Gugh is attached to St Agnes by The Bar, a low ridge of sand that is uncovered at low tide. Areas of blown sand occur in low-lying parts of Gugh and in certain places on St Agnes, around Wingletang Down in the south and Big Pool in the north. On Wingletang Down, there is also a good example of an area of maritime or 'waved' heath, where *ling* heather has been eroded and 'pruned' by the wind. The island has its fair share of rocks weathered worn into shapely forms; look out for the Punch Bowl near Beady Pool and The Nag's Head above St Warna's Cove.

Pottery evidence suggests that St Agnes may have been settled in Neolithic times and additional indications of this can be seen in the remains of a rectilinear field system visible on the downland above Long Point. Remains of prehistoric and Romano-British houses have been found at Porth Killier, around Higher Town and on the lower slopes of Kittern Hill on Gugh. There are a number of entrance graves on Gugh, Obadiah's Barrow being particularly impressive with four of its capstones still intact; excavations of this grave revealed human remains and funeral urns. Nearby is a Bronze Age menhir or standing stone over 2 metres high called 'The Old Man of Gugh', which is a prominent feature on the island's skyline. Back on St Agnes, 43 cairns survive on the moorland around Wingletang Down in the form of low circular mounds, many covered by *gorse*; these are indicative of a prehistoric cemetery. Nearby on the western side is St Warna's Well, a place of unknown antiquity where three stone steps lead down to a small underground chamber covered by a single capstone. Here is the site of a holy well dedicated to St Warna, the patron saint of shipwrecks, who, legend has it, landed in her coracle from Ireland at the nearby cove of the same name. If only the stones could speak, what tales would be told of wrecks and smuggling and of the men of St Agnes, who were some of the finest and bravest pilots of all the islands.

Farming has always been important to the people of St Agnes, but judging by ancient field patterns, the area of farmland that is enclosed has shrunk by almost a half since medieval times. This now essentially comprises a mixture of bulb strips on the hill slopes with intervening areas of grass fields enclosed by stone walls. The bulb strips are bordered by shelter hedges, which include a wide variety of species such as *tamarisk, euonymus, veronica* and *pittosporum*. Stunted *elms* line many of the narrow lanes between the settlements and these, together with the shelter hedges, give almost a wooded appearance to the island, particularly when viewed from the lower slopes near the coasts.

The lighthouse, one of the oldest in the British Isles, is situated on the central ridge at Middle Town and dominates not only St Agnes itself, but also the rocky islands all around, particularly those out to the west. As we have seen, ships of many kinds and from many countries have foundered on the treacherous reefs of Scilly, and old-established trading companies, including the East India Company, had pleaded for the building of a light on Scilly. So in 1680 the highest point of St Agnes was chosen by Trinity House as the site of a lighthouse; it was built and painted white to serve as a day mark and stood just over 22 metres tall. The first light consisted of a coal fire lit in an iron brazier inside the lantern, and although it burned brightly, at times it needed constant attention and was unreliable. So in 1790 the coal-burning light was replaced with copper oil lamps with 21 revolving reflectors. Eventually, in 1911 this, the oldest light in Scilly, was discontinued when the automatic Peninnis light on St Mary's was built. It is interesting to note that the construction of the St Agnes Light was not without opposition. Officials from the Isle of Wight argued that they would suffer from a loss of revenues because many more ships would favour Scilly for victualling, and the Governor on St Mary's complained on behalf of the Scillonians that they would be subject to a loss of profits from wrecks! In those days too it was said that when times were hard the inhabitants turned to St Warna for help in sending them wrecks to relieve their suffering!

Near the cliff edge on the coastal path to the south-west of the old lighthouse is a group of small pebbles laid on the ground in the form of circular paths. This is called the Troy Town Maze and is thought to have been constructed by a bored St Agnes Lighthouse keeper in 1729 to while away the time when there was nothing to do. In fact it is not really a maze, where there is a choice of turnings; it is a labyrinth, strictly speaking, with a set route to be followed. There are a number of these mazes on the islands but the Troy Town Maze is the oldest. Some say there may have been an earlier origin to the maze but nobody is really sure when and for what reason it was built. Another mystery on these islands of mysteries!

St Agnes is home to many of the birds already mentioned, and because of its variety of tall hedges and shrubs it is also the place to see nesting birds such as the *chaffinch* and *willow warbler*.

The chaffinch, about 15 centimetres long, is easily recognisable with its bold white shoulder patches and, in flight by its white outer tail feathers. Seen close at hand, the male is a truly handsome bird with a lovely pale pink breast, warm pinkish-brown back set off with a greenish rump and a slate-blue crown and nape. The female is less well adorned, with the same white markings but generally of a dull grey colour and none of the smartness of her mate. As with most birds in winter, the male is much less bright but still distinctive. In early spring chaffinches pair up and the couples seek out a territory in a hedgerow or area of scrub. The cock bird sits on top of a branch or bush and proclaims his right over his territory with a joyous ringing trill ending in a flourishing 'choo-ee-o' whenever anybody passes by. The hen bird does most of the work in building the nest in the branches of small trees in the shelter hedges, about 1 to 4 metres off the ground. The compact cup-shaped nest is made of moss, wool and grass, all beautifully felted together with cobwebs and lined with feathers and down. To help conceal it, the nest is finished off with a decoration of grey lichens on the outside. The four to six greenish-blue eggs, spotted and streaked with reddish-brown, are laid from early April, with second broods starting as late as June. The hen bird alone incubates the eggs but the male is a most attentive partner, feeding her when on the nest, and bringing a plentiful supply of caterpillars to thrust down the ever-ready open beaks of the young when they hatch.

As autumn approaches and the leaves from the wind-blown elms and other deciduous bushes and trees begin to fall, so the chaffinches begin to congregate in small flocks, generally in single-sex groupings. Some of the males actually migrate south to the warmer climate of the Continent before returning in the spring. The chaffinch is the commonest of British birds and you will have no difficulty in seeing them on the inhabited islands of Scilly where there is cover and shelter; in winter the birds will be seen out in the bulb fields in their flocks, feeding mainly on weed seeds and insects.

The willow warbler is a migrant and summer visitor but does not breed in large numbers on the islands. A small greenish-yellow bird with brown legs, the willow warbler arrives in late March from central Africa, departing again in October, another wonder of migration when one considers this small bird making so long a journey at a never-faltering speed of some 40 kilometres an hour. I remember watching a newly arrived willow warbler singing its heart out on a still spring day amongst the tangle of bushes and hedges just to the north of the lighthouse. What a thrill it was just to sit beside the wall looking out to the west at the magnificence of the rocky islets set against a background of blue sea and sky; and in my ears the song of this delightful bird, a rippling sequence of liquid notes, beginning faintly then rising strongly, only to fade again. I am sure it had its mate nearby; if so, it would certainly soon begin to build a nest. This would be well-hidden in a tuft of grass and dead bracken amid low bushes such as bramble. The nest itself is dome-shaped, made up of grass, bracken, moss and dead leaves, and is lined with hair and feathers. Often it is well sunk into the ground and only the oval entrance (about 6 centimetres wide) is visible. The willow warbler lays four to six white eggs, spotted with pink or red, in April or May, with repeat clutches until July. The nest, although very well concealed, can be found by watching the parent birds when they are building or when feeding their young; but be very careful and do not linger and do not disturb the nest itself, or the birds could desert and not return.

Another migrant warbler, the *chiffchaff* is very similar to the willow warbler, not only in its migration route and nesting habits, but also in its size and appearance. However, it can be recognised by its blackish legs and by its distinctive song, sung from the top of a tree, a deliberately repeated series of notes 'chiff, chaff, chiff, chaff' ...

Gugh provides the nesting sites for a number of *common shell-duck*, which we mentioned earlier in the chapter on Tresco. This is a very handsome species of bird with both the male and female having a similar appearance. Each has a white body, the front of which has a chestnut-coloured belt encircling it, and each has some black on the back and a greenish-black head. The male can be distinguished from the female by the prominent knob on its red bill and by the fact that it is bigger, some 60 centimetres long.

The common shellduck first nested in Scilly in 1958 and since then its numbers have increased slightly until now there are a pair or so on Samson, Gugh and St Agnes. It is a migrant duck that sometimes stays on through the winter, but just as likely it roams around the shores of the Continent when not actually on the islands breeding. In early spring the birds pair off and look for likely nest sites; a favourite site is a rabbit hole concealed under a bramble or gorse bush but sometimes it is up to a metre or so deep under a large rock. The nest itself is almost entirely made of the female's own pale grey down, together with some vegetable material such as dry grass and bracken. The 6 to 12 pale cream-coloured eggs are laid in April, with possible repeat clutches in May or June. The exact locality of the nest is often given away by the presence of the male bird standing on guard nearby, the female doing all the work of incubation herself. As soon as the ducklings hatch out, both adults escort the brood to the water; for some reason they all make their way across St Mary's Sound and The Road to the beaches of Tresco some 6 kilometres distant. What a sight they make as the black-spotted white ducklings take to the sea behind their brightly coloured parents. As soon as they reach Tresco they all make for one of the large expanses of inland water, usually Great Pool. Clearly they prefer these placid waters to the beaches of St Agnes or Samson, where the Atlantic rollers fling themselves onto the black rocks in almost unceasing strife.

The common shellduck is usually a very sociable and gregarious bird, except at breeding time, and as the young ducks grow bigger so they join up with other youngsters to form quite large groups on the water, generally under the guardianship of one of the adults. Their life is idyllic, with few natural predators around and plenty of food in the form of small aquatic creatures and vegetation. As autumn approaches and the first real winter storms hurl themselves across the beaches, so they become restless and some begin to leave their island home to seek the wider shores of the distant Continent. Many, of course, stay to enjoy the mild climate, and these are joined by other winter migrants from the mainland.

Like so many of the islands, St Agnes is a wonderful place to see overwintering and migrant birds, which begin to arrive from late September onwards. Sit beside the rushes at Big Pool as the green leaves of summer just begin their subtle change of colour before falling to the ground. The last *swallows* are beginning to congregate in preparation for their long flight south, swooping low over the water to feed on the plentiful insects. Some of the first waders soon arrive to spend the winter; *lapwings*, with their erratic flight pattern and slow flapping wing-beats, showing off their black and white plumage; and *golden plovers*, pecking away at the pool edges, their golden speckled plumage blending in well with their surroundings. Just behind you and over the stony bank a small flock of *sanderlings* run in and out of the sea as the waves break on the sandy shore at low tide. They fly off together and then land as one, a short distance away. These grey and white birds have come all the way from northern Scandinavia and Greenland and are very tame on Scilly, probably having never seen humans before. *Turnstones* too are common, turning over stones and shells with their stout pointed bill, looking for food and occasionally uttering a staccato call, 'tuk-a-tuk', and a long rapid whistle. The turnstone is some 24 centimetres long, with a winter plumage of dusky brown and a white throat and distinctive short orange legs.

As you wander along the little lanes of St Agnes, look out for other autumn migrants such as the *blackcap, pied wagtail* and *redpoll*.

I don't know whether the *hedgehog* has yet reached St Agnes, but it is worth a mention here just in case it has found its way over from St Mary's. There were no hedgehogs in Scilly before about 1984 and then they appeared on St Mary's, certainly introduced by someone, and now their numbers have increased rapidly. The hedgehog lives in fields, hedgerows and gardens, coming out at night to feed, mainly on insects, snails, slugs, worms and fallen fruit. It will overturn small stones to search for food and in the spring and early summer it is a voracious feeder on birds' eggs, and here lies the problem. As we have seen, many of the birds on Scilly are ground nesting, and some, like the *ringed plover* and *oystercatcher*, could be decimated if hedgehogs spread to other islands and their numbers were allowed to get out of control.

Hedgehogs mate in early spring and summer, and up to seven young are born from May to September. The nest of grass and leaves is generally built under a tree root, or in a disused rabbit hole, of which there are many in Scilly. The natural enemies of the hedgehog are the fox and badger, but since there are none of these on the islands, this well-loved and popular creature will be able to multiply quickly. Only on St Mary's is it likely to come to a sticky end beneath the wheels of traffic.

The story of the hedgehog in the Outer Hebrides of Scotland should be a salutary warning to those who concern themselves with conservation and protection of the natural environment on The Isles of Scilly. In 1974 a total of four of these animals were imported to the island of South Uist to control slugs. Now there are estimated to be at least 10,000 hedgehogs on the island; a multiplying on a Malthusian scale. By 1993 the creatures had spread throughout South Uist and into Benbecula and North Uist via the causeway linking the three islands. The hedgehog has caused a good deal of harm to nesting birds, such as the lapwing, dunlin, oystercatcher, ringed plover and the rare corncrake; a recent survey suggests that up to ten per cent of nests have been destroyed. Radio tagging of individual animals clearly showed that they were wandering over the foreshores and sandy grasslands and devouring many birds' eggs. Urgent action is now being taken to reduce the threat to the bird population by removing the alien invader. Perhaps steps should be taken now on Scilly to reduce the hedgehog population before the problem gets out of hand.

Like the other islands, St Agnes and Gugh are home for a wide variety of wild plants. *Wild thyme, wild mignonette* and *viper's bugloss*, common plants on the mainland, have one of their few localities in Scilly on Gugh and St Agnes.

Wild thyme has a creeping stem, from which arise the flowering shoots containing masses of tiny, two-lipped, rose-purple flowers which appear from May to September. The flowers are full of nectar and highly fragrant, and consequently attract many types of insect. In spite of its diminutive nature, it is really a shrub, with a woody rootstock that penetrates deeply. Strangely for a plant that is so common in Cornwall, it only grows on Gugh and on Great Ganilly in the Eastern Islands. You will see it at its best in late June when it carpets the ground with purple blooms.

Wild mignonette is a very rare plant on Scilly and can only be found now on some waste ground near The Bar. About 60 centimetres tall, this bushy, hairless plant has deeply divided leaves and small six-petalled yellow flowers carried in spike-like clusters at the end of the stems. It flowers from June to September. Wild mignonette once grew on the sandy ground at Hugh Town, St Mary's, but this colony was destroyed in 1939. It was probably brought in by shipping from the Mediterranean many, many years ago, having been first recorded in 1864.

Viper's bugloss grows up to a metre tall and has stout stems, bristling with hairs, which carry attractive, bell-shaped, purple and blue flowers from June to August. After pollination, each flower produces four seeds, which are said to resemble a viper's head, hence its name; for this reason also it was once used as an antidote for snakebite. With this background, it is not surprising that it was also recommended as a cure for melancholy. Like the previous species, it can only be found growing in wasteland just above The Bar. Viper's bugloss is an allied species to the huge *tree echiums* which are found as garden escapes on many of the inhabited islands and with which it should not be confused.

On the sand and shingly beaches, particularly at Periglis Bay, look out for the low, glossy, green cushions of *sea purslane* also called *Sea Sandwort*. The fleshy oval leaves are arranged in pairs, each pair at right angles to those above and below it. The somewhat inconspicuous five-petalled, small greenish-white flowers have short stalks and expand only in the morning sunshine from May to September. On the mainland costs of Yorkshire the leaves and stems of this plant were once used to make a pickle.

Another plant of the seashore, particularly where the sand is fine, is the *prickly saltwort*. Its pale, blue-green fleshy leaves are stalkless and have a sharp spine at the tip; the tiny green flowers appear in late summer at the base of the leaves. This plant thrives in salty conditions and is often partly buried in blown sand. In Mediterranean countries it was once dried like hay and then burnt to produce a semi-fluid alkaline matter which was used as washing soda, hence its Latin name *kali*, from which the word alkali is derived. I wonder whether it was used on Scilly for this purpose many years ago. Look for it on Dropnose Porth on Gugh, but it can be found on many of Scilly's sandy beaches.

One plant you may be able to spot on your walks around the island is *bear's breeches*. This stout, hairless plant up to 150 centimetres tall has its origins in the western Mediterranean and Portugal. It was probably introduced to Scilly as an exotic garden plant and then escaped or was deliberately planted. The large two-lipped purple and white flowers are held in a dense terminal spike and are very distinctive. The plants are not easily seen from roads or public footpaths, but on St Mary's and Tresco they are much more obvious; look for it on field banks near to the shore. The colony on St Agnes was recorded as long ago as 1800, when it attracted great interest from the early botanists.

Another distinctive and attractive plant to be found just about anywhere on St Agnes is the *great mullein*. It is not a common plant on Scilly and although widespread on roadsides, walls and dunes, you will find only a very few plants in each locality. This is a tall, woolly plant some 120 centimetres high, the flower spike of which carries many clusters of yellow flowers, which bloom from June to August. The leaves at its base are very large, sword-shaped and covered with soft, wool-like hairs. Our Saxon forefathers dipped the dried stems and flower heads in melted grease and then used them as torches at medieval church festivals. I wonder if they were ever used in this way in the ancient places of worship on Scilly. Before the introduction of cotton, the leaves and stem were used as lamp wicks. The leaves are also used in homeopathic medicine as a tincture to ease coughing. The whole plant, including the flowers, has soothing properties and has been used over the centuries in many medicinal concoctions to ease such ailments as asthma, nervousness, neuralgia and stomach cramp.

In the autumn, look out for that lovely little orchid with its distinct spiral of white flowers, *autumn lady's-tresses*, which we mentioned before in Chapters 2 and 7. You will find it in the short grass around Troy Town and on Gugh.

Big Pool is the place to see dragonflies. There are only two species of dragonfly on Scilly, the *common sympetrum* and the *common ischnura*, and both can be found around the fresh water pools on St Mary's and Tresco, and here on St Agnes. Dragonflies have existed since the Carboniferous era of 350 million years ago, when they had wingspans of some 60 centimetres or more. Not only are they one of the oldest of the world's insects, but they are also one of the fastest, with flight speeds of 100 kilometres an hour. They feed on flying insects up to the size of small butterflies, which they catch and eat in flight. Sometimes they perch on a fence post or branch to eat the larger prey. Often you will see two dragonflies flying together 'in tandem' in a mating embrace. The female lays her eggs in the water and they hatch into nymphs or larvae. These nymphs live in the water for up to two years, living off anything from tiny insects to large tadpoles, depending on the size of the nymph itself. When mature, the nymphs leave the water by climbing up plant stems in early summer. They then shed their skins and emerge as winged adults, which only live for about a month.

The male common sympetrum is easily recognisable because of the distinctive red colour of its body, which is about 4 centimetres in length; the female being less obvious with a yellow-brown body colour. The common ischnura is, strictly speaking, a damselfly, which is a small relative of the dragonfly. It is about 3 centimetres long, with a thin black body which has a blue segment at the tip. Both sexes are similar. Damselflies can be easily distinguished from their larger cousins the dragonflies by their weak fluttering flight and by the fact that they usually rest with their wings folded above their body; dragonflies rest with their wings spread horizontally.

Sit and enjoy the spectacle of these lovely flying insects with their transparent wings as they dart around the pool with great energy and zest.

Butterflies, which are the descendants from the great dragonflies of the past, are abundant on St Agnes, as everywhere else on the islands. Look out for the *small copper* as it flits from flower to flower. It is a very attractive butterfly, whose wings are a bright copper colour with black markings. Occasionally you will see some with a row of small blue spots across the hind wings. It is often seen basking in the sunshine with its wings spread open but it is quick to fly off and intercept other butterflies that intrude on its territory. Each butterfly lives for about a month, so there is not much time before the female must lay her eggs to perpetuate the species. She lays her eggs singly on the upper surface of the leaves of the *sheep's sorrel* plant, a small member of the Dock family that is abundant on all the islands. These hatch out in six days and the first generation of caterpillars feed for about three weeks before pupating and forming chrysalids. These chrysalids last four weeks, producing the second generation of butterflies, which fly in late summer and early autumn. The caterpillars of these may live for seven months or so, hibernating over winter to produce the chrysalids from which the new generation of adult butterflies emerge in the early spring. The small hairy caterpillars of this butterfly are well camouflaged, the early ones being green and the later ones being green and pink to match the change in colour of the host plant in late summer.

So we leave the flying insects and for a moment, as we look out at the vastness of the ocean, we turn briefly to some of the larger creatures to be found in its reaches such as the *basking shark* and the *common dolphin*.

Occasionally as you travel between the islands, in the deeper waters you may catch a glimpse of the large dorsal fin of the basking shark as it cuts through the water, and if it is calm you will be able to see its gaping wide mouth and huge body just beneath the surface. This is the second largest species of fish in the world, sometimes over 10 metres long. The basking sharks feed on tiny sea creatures called plankton, and as they cruise very slowly (about 2 knots) in the water with their mouths open, they filter large quantities of water through their gills. These enormous gills are equipped with fine networks of interlocking combs which filter out the tiny plankton, which are then digested. The basking sharks are migratory and spend the summer in the waters off the west coast of Scotland and Ireland, but little is known of their exact movements and you could see them just about any month in the seas around Scilly.

The common dolphin is one of the most beautiful of animals to watch as schools of them glide seemingly effortlessly through the blue waters beside or just in front of your boat. Watch them, as in graceful curves they leap free of the water in sheer abandon, to plunge again, leaving hardly a splash. They are reasonably common around Scilly and they feed mainly on fish such as *mackerel* and *herring*. Sadly, quite a few get caught up in the nets of fishermen and in 1989 I found quite a number on the beaches, entangled in green and blue netting.

Very occasionally you may be lucky to come across one of the truly exotic creatures of this planet, either in the water itself or more likely washed up on the beach. Over the years the island people have recorded such 'animals' as the *leathery turtle, electric ray, blue marlin*, and even 'sea serpents', which were probably *oarfish*, eel-like creatures that can grow up to some 15 metres long. Keep your eyes open wherever you are; you never know what will appear when you least expect it.

Leaving the calm and solitude of St Agnes and crossing St Mary's Sound, you may be lucky enough to see one of the annual Gig Races in progress. The slim high-pointed bows of the six brightly painted gigs cut through the blue water and the bare backs of the gigs' crews glisten in the afternoon sunlight as they strain to get maximum power out of their oars. Gleaming white spectators' launches hold position just behind the racing boats and yells of encouragement sweep across the sea, seemingly making little difference to the labouring oarsmen. The lead keeps changing with surprising rapidity but gradually the dark blue boat edges ahead until at the first marker at Steval Rock, a yellow gig cuts inside the turn and races ahead, never to lose the lead again. The crews never admitting defeat keep rowing hard until they all finish within minutes of each other at Rat Island quay. So the race ends with all the crews enjoying the comradeship and shared pleasures of pints of refreshing beer. At the same time our boat enters the harbour and we prepare for our exploration of the last island on our journey through The Sun Islands.

The Old Man of Gugh

Nag's Head   *Punch Bowl*

Sir Cloudesley Shovell's Monument at Porth Hellick and Loaded Camel

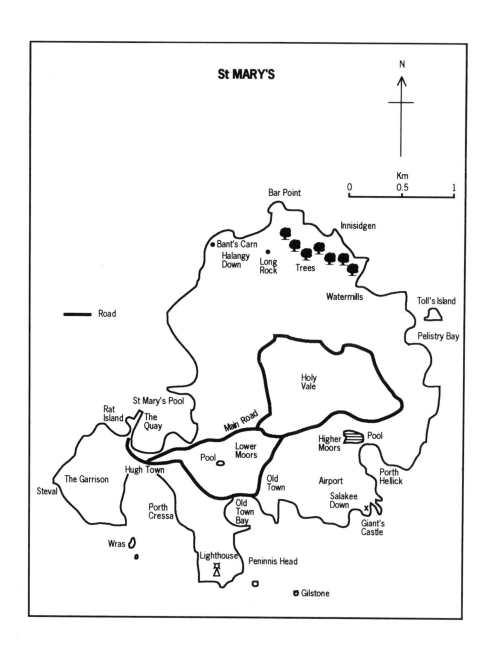

11

ST MARY'S

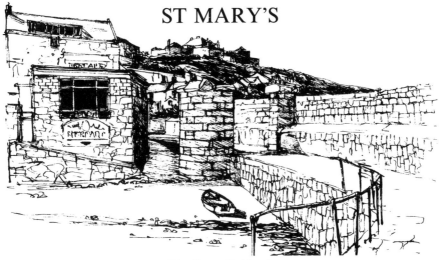

The Quay, St Mary's

St Mary's, approximately 4 kilometres by 3 kilometres in size, is by far the largest of the islands, and has some 1,600 people, seven-eighths of the total population. The island consists of two unequal but relatively high parts joined by a narrow sandbar on which is situated the small settlement of Hugh Town. This settlement, vibrant with life and colour, is the centre of commercial activity on the islands and incorporates the islands' council offices and all the other necessities of a civilised society. Shops, cars, aircraft and many people, both visitors and Scillonians, are to be found on St Mary's, but it also has much to offer the reader of this book. There are some wonderful nature reserves and numerous places of restful tranquillity. Everywhere there is evidence of its ancient history where one can marvel and reflect on those people of long ago who travailed and sustained themselves here on this island outpost, remote from a mainland and continent they never knew.

Rising from a low rocky coastline, the interior of the island is gently undulating and includes the rounded plateau hills at Halangy Down, the golf course and airport, the small wooded valleys of Watermills and Holy Vale, and the broad, low-lying damp areas of Higher Moors and Lower Moors, each with their freshwater pools. The island is almost entirely made of granite, but there are some fine areas of blown sand, creating small sandy bays, especially around St Mary's Pool, Bar Point and Porth Hellick. St Mary's has the appearance of a wooded landscape, for in the mid-nineteenth century many *Monterey pine* trees were planted as shelter belts and a large number of *elm* trees were planted as hedges. These elms have now grown to form lines of wind-stunted trees along field boundaries and beside the narrow winding lanes. So the island is one of contrast, where at every turn you will come across something different, and because of this it has a good cross-section of wildlife.

Most of the recorded prehistoric settlements, field systems and entrance graves are to be found around the island's present coastal fringe. The main settlement in medieval times was Old Town, lying on the south side of the island, the more modern Hugh Town having been developed during the seventeenth century after the construction of Star Castle in 1593.

The most famous of the Scillonian entrance graves is to be found at Bant's Carn, some 2 kilometres north of Hugh Town, which when excavated in 1900 revealed four piles of cremated bone as well as Bronze Age pottery. A substantial field system of indeterminate age lies nearby. Other prehistoric grave sites can be seen at Buzza Hill (above Hugh Town), Innisidgen (on the north-east coast), and at Porth Hellick Down (on the south-east coast). A standing stone, or menhir, called Long Rock, over 2 metres high, can be seen just to the east of Halangy Down. It is apparently still in its original position, leaning towards the south-east; interestingly, flint tools have been found nearby. All this evidence suggests that St Mary's was the centre of population those many, many years ago in prehistory, as it still is today. In recorded history, of course, St Mary's has been the focal point of the construction of fortifications which has continued through the centuries right up to modern times. In a book of this nature it would be repetitious and dull to record them all, but from an ancient cliff castle at Salakee Down to World War II concrete machine-gun posts at The Garrison, the marks of man and his need to defend himself can be seen everywhere on the island. So too can more practical historic artefacts, such as the fish-salting trough on the east side of Old Town Bay. This is a deep rectangular trough, cut from a single block of granite, which was used for salting fish at a time many hundreds of years ago when all the fish of the islands were brought to Old Town for curing. The fish were laid out on stages in a nearby field to dry in the sun. Sadly the trough is now neglected with debris of all kinds inside and around it, chains, old pieces of netting, drain pipes and worn-out tyres.

The granite rocks of Scilly are renowned for their weird and strange shapes and there is no place better to see them than at Peninnis Head. Here, on this utterly exposed place, the rain, wind and salt spray of the Atlantic gales have weathered the granite into rocks of fantastic shapes, which over the years have been given such names as Kettle and Pan Rocks, Tooth Rock, The Pulpit and Monk's Cowl. Not far from the monument of Sir Cloudesley Shovell at nearby Porth Hellick stands the rocky outcrop shaped and known as The Loaded Camel. Most of the granite on Scilly has geological fault lines which formed thousands of millions of years ago when the earth came into existence. Some of these lines are vertical faults which weather into columnar shapes, like the Tooth Rock; and others, like The Pulpit, have horizontal faults where erosion has produced flat slabs, often balanced upon smaller rocks. Sometimes where both vertical and horizontal faults occur, decomposition of the granite progresses until just a chaotic heap of rocks of all shapes and sizes is formed; such a sight can be seen at the tip of Peninnis Head itself. If you sit on the grassy slopes just south of Peninnis Lighthouse, look out at these grey rocks with their belichened colours of golden ochre and raw sienna, and see if you can spot two other faces that are well worth finding: the Laughing Man and the Old Witch. As the sea crashes up against the great granite bastions with spray flung high in the sighing wind, try to imagine the aeons of time that have passed in order to form these magnificent sculptured stones. What tales they could tell if they were less than inanimate; of shipwrecks and of pirates, of peoples, their loves and hates, their feelings and their reactions to battles of long ago. Superimposed forever on this landscape are the never-ending noises of the restless sea in storm and in calm. As you sit alone amongst all this splendour your mind begins to wander as the hot sun and alluring smells of the sea-shore and countless wild flowers dull the senses, but the sound of an aircraft taking off nearby brings you back to reality.

St Mary's has its complement of nesting seabirds and waders, such as the *black-backed gulls* and the occasional *ringed plover*, but it is the common garden birds that you are more likely to see as you walk along the leafy lanes; the *robin* and the *wood pigeon* are two species that will soon make themselves known.

The robin, too well known for me to describe in detail, is a lover of thickets and bushy places and is especially at home in a small garden. Robins are quarrelsome little birds that are seldom seen

135

together outside the breeding season. Both the cock and hen bird sing, and their voice, a persistent and often rapidly repeated 'tic' or 'tsissip', is heard practically the whole year round. Their song, too, can be heard throughout the year and a sweet and pleasant song it is with its deliberate series of short warbling phrases of fine quality. In the autumn the song becomes particularly plaintive and melancholy in its quality, perhaps anticipating the onset of winter with its storms and rain.

Robins are classic defenders of their territory, which they have delineated for themselves after their summer moult, and they will engage in fighting any other robins who dare to intrude. The birds will usually pair up from mid-December onwards, when they posture and sing loudly at each other; the hen quickly learns the territorial boundaries of her chosen mate by being chased back over them by neighbours. Other birds, such as the *dunnock*, are chased away by the robin's threat display, which consists of puffing out the red breast with a swaying movement and outstretched neck. The hen robin alone builds the nest of roots, dead leaves and moss, lined with hair, in just about any conceivable situation – holes in banks, amongst ivy growing on walls and trees, in flower pots, old kettles, rabbit holes and many other odd places. The five or six white eggs, freckled with dull light red, are laid from March, with repeat clutches until June. Juvenile birds leave the nest in a spotted brown plumage not unlike miniature thrushes, but by the time autumn approaches they are difficult to distinguish from the adults with their smart red breasts. Once the young are independent and feeding on insects, worms and berries, then the parents and young separate, each taking up its own territory and singing a song of warning to others to keep their distance; and so the cycle of nature is ready to repeat itself ad infinitum.

As you walk under the large Monterey pines or along the leafy lanes in early summer, the distinctive sound of the wood pigeon with its five-note cry, 'coooooo-coo, coo-coo, coo', echoes through the branches. The largest of the five species of pigeon in this country, the wood pigeon is easily recognised by the broad white patches on the side of its neck and white band across its wings. Like others of its species, it feeds on cultivated crops such as grain, cabbages and young seedlings, and can be very destructive to the farmer. Its nest is a lattice of small twigs woven into a loose platform placed in the branches of bushes or trees some 2 to 20 metres above the ground. The nest is seldom well concealed and can easily be seen through the branches; the two white eggs are laid any time between March and September. The young hatch out after about 17 days and are fed by both adults with a protein-rich, cheese-like 'milk' produced in the bird's crop. The wood pigeon is the only bird that produces such a milk, which is similar to that of a mammal. No sooner has the first brood flown off the nest than another is started and three broods are quite normal for the species.

Autumn finds the wood pigeon gathering in quite large flocks, the resident birds being joined by others from the mainland, and it is at this time that it does most of the damage to crops. John Masefield's description of the bird in his poem 'Wood-pigeons' is most apt:

> It is a beauty none but autumn has,
> These drifts of blue-grey birds whom Nature binds
> Into communities of single minds
> From early leaf-fall until Candlemas.

Another common breeding species of pigeon in Scilly is the *collared dove*. Unknown in Britain before 1952, this bird now breeds throughout the country and its call, 'coo, coo-o, cuk', accentuated on the second syllable, is now a familiar sound. It is easily recognisable by the black collar at the back of its neck and the white beneath the end of its tail. Like the wood pigeon, it nests in trees and lays two white eggs between March and September.

One of the lovely sounds of the open country on St Mary's in the spring and early summer is the song of the *skylark* as it hangs in the air on fluttering wings and pours out its liquid notes. On the ground this bird is inconspicuous, with its dull brown plumage and short, rounded crest, but in the air, even as a mere speck against the blue dome of the heavens, it really becomes something special. Its musical outpouring, most often heard from March until July, is long and sustained as it soars in the sky. As it circles over its nest site it continues to sing, until after a few minutes it suddenly closes its wings and drops to the earth like a stone, just preventing a crash landing by opening its wings at the last moment. The skylark builds its nest of grass on the ground in crops or under a tuft of vegetation; the three to five dark brown speckled eggs are laid from April to July and two broods are common. The incubating female sits quite tightly on the nest, so if a small brown bird flies up suddenly from close to your feet, stop and look very carefully in the grass around you. The skylark is basically a ground-living bird that feeds on seeds and insects, and because of this and its nesting habits, it is very vulnerable to intensive farming. On the mainland its numbers have been drastically reduced by over-grazing of its habitat, by pesticides in the food chain and by herbicides that reduce the amount of weed seeds available as food for its young. None of these apply in these beautiful islands and the only restriction on its numbers is the amount of wide open space which it needs to survive. At present there are just two or three pairs of skylarks breeding on St Mary's, its only locality.

St Mary's is an island with a rich variety of wild flowers. In spring there are fields covered with *Bermuda buttercups* giving a brilliant colour of lemon-yellow to the gentle countryside. In summer as you walk along the beaches to the east of Hugh Town, the tall stout stems and rich purple flowers of *tree mallow* (described in Chapter 9) stand out amongst the scattered debris of lobster pots, fishing nets and bric-a-brac associated with the repair of boats. Two other members of the Mallow family grow here, the *common mallow* and the *cretan mallow*.

The common mallow is a plant of considerable beauty, with a hairy erect stem growing to about half a metre high, and large pale purple flowers with darker lines that converge to the centre. The five petals, each nearly 2 centimetres long, are heart-shaped and are to be seen from May until September. The leaves of the common mallow are kidney-shaped, with edges broken up into five lobes which have toothed borders. The common mallow can be distinguished from the tree mallow because it is smaller and more straggling in its nature and has a coarser and more hairy texture. Look for it on waste ground, roadsides and about buildings; it is more likely to be found inland and away from the shore than either the tree mallow or the next species, the cretan mallow.

The cretan mallow is one of the real delights of The Isles of Scilly, being more abundant here than anywhere else in Britain. This species is a native of the Mediterranean region and the coast of France and was first discovered on Scilly in July 1873. Its appearance is very dependent on seasonal climatic conditions and germinates readily in warm wet springs, so you will find it in flower at any time between March and September. Its flowers are smaller and paler than the common mallow and its petals are almost rectangular in shape, often with almost parallel sides; its foliage too is smoother than that of the common mallow. In Scilly it seems to be confined to three islands, St Mary's, St Agnes and Tresco, where it is restricted to disturbed ground; for some reason it has not spread to any of the other islands.

The Latin name of the mallows, *malva*, derives from the Greek *malache*, meaning soft, and the mucilaginous juices of these plants have been used medicinally since 700 BC. They have more soothing properties than any other plant and have been used as an intestinal stimulant and as a treatment for digestive and urinary disorders. In the twelfth century mallow was widely recommended as a treatment for such maladies as drowsiness, headaches, kidney diseases and poisoning. The boiled leaves of the common mallow can be used as a vegetable and the flower centres, sometimes called 'fairy cheeses', can be eaten raw. Like so many others, this was an important plant which had wide uses over the centuries.

As you walk eastwards along Telegraph Road, look for the sign indicating the Lower Moors Nature Trail. Here is a place of beauty and serenity quite different from the sandy, rocky shores and heather-covered hills of the other islands; for in this place are streams and pools of still water with reeds and shady willows of various species. The main willow species is the *grey sallow*, with oval leaves which are a dull green above and paler below with prominent veins. It grows to some 5 metres high and flowers very early; its silky silver-grey male catkin buds open in January giving a feast of nectar to early bees and other insects. The leaves and fruits of this sallow reach maturity at the same time as the male catkins wither and fall when their pollen is shed. The female ones, however, remain attached to the branches, finally bursting into pure white masses of fluff, which, along with their attached seeds, are scattered far and wide by the wind. Many of the willows were planted in Scilly because the branches were useful for making baskets for lobsters and crabs; interestingly, Porth Hellick on St Mary's is derived from old Celtic words meaning 'cove of the willows', so there is little doubt that it is truly a native tree.

In midsummer the sides of the trail are adorned with metre-long and gracefully arching lance-shaped fronds of the *male fern*. Extracts from the roots of this fern are used in homeopathy to treat septic wounds, ulcers and varicose veins. In ancient Greece these extracts were considered to be a valuable vermicide, and even today they are used by veterinary surgeons for expelling tapeworms. Nearly all ferns reproduce by means of spores which appear in late summer and autumn as brown patches on the back of the fronds; these are dispersed by wind and animals.

Ferns are among the most primitive plants on earth, and they date back more than 300 million years, when they were generally much larger than they are today. In fact, the world's first forests were formed from tree ferns reaching 30 metres high, and these form the basis of our present-day coal deposits. One of the descendants of these ancient plants is the enormous *Royal fern*, which is one of the great attractions of this small nature reserve on St Mary's. This fern is aptly named because it grows to a height of over 3 metres and has fronds that are cut into broad elegant leaflets. In some of the fronds the leafy portion is greatly reduced and is densely coated with reddish-brown spore capsules. These spore-rich fronds form into large spikes, similar in appearance to the clustered red flowers of the Dock family, but of course the similarity is very superficial as ferns have no flowers. You will not fail to see these tall brown spikes of the Royal fern sticking up above the aquatic vegetation in June as this plant loves to be deep-rooted into wet boggy ground.

Look at those deep pink flowers of *ragged robin* and the tall robust stems of the *hemlock water-dropwort* growing in the wet ditches.

Ragged robin, which blooms in May and June, is a descriptive name for this plant, which has unmistakable rose-red flowers, each with five petals deeply cut into four lobes, giving them an overall 'ragged' appearance. The slender, reddish stem is about 60 centimetres tall, with lance-shaped leaves arranged in pairs along its length. Its Latin name, *flos-cuculi*, reflects the fact that its appearance coincides with the arrival of the cuckoo and its calling. Folklore associates plants with 'robin' in their name with goblins and evil, and it is still considered unlucky to pick this plant and bring it indoors.

As its name suggests, the hemlock water-dropwort is one of our most poisonous species of plant, and on the mainland it has caused the deaths of several people recently, particularly amongst tourists from the Continent who have mistaken it for wild celery. On Higher and Lower Moors it grows abundantly along the stream sides in July and August. Its hollow, hairless stems support large umbels of small white flowers which have a pleasant parsley-like scent.

The only marsh orchid found in Scilly is the *southern marsh orchid* and this can be seen flowering in June in the damp meadows on Lower Moors; it has purplish spikes some 40 centimetres tall. The stem is hollow and the dark green, lance-shaped leaves are generally unspotted, unlike those of some of the other marsh orchid species.

In the late summer the large white trumpet-like flowers of the *hedge bindweed* are borne on long trailing stems that scramble over the taller vegetation, not only here on Lower Moors, but just about anywhere on the island. Occasionally the flowers are a handsome pink colour and what a sight it is to see one of these with a *convolvulous hawkmoth*, as described in Chapter 5, feeding on its nectar.

You will see wild flowers on Lower Moors just about every month of the year. In early spring the earth banks are adorned with the yellow of the *lesser celandine* and soon the ditches themselves are speckled with the yellow of *lesser spearwort*. Birds of many kinds visit the reserve and the dawn chorus of late spring is filled with the song of *blackbirds, song thrushes* and a great many of our *warblers. Moorhens* and *mallard* ducks breed around the pools and sometimes you may catch a glimpse from the bird hide of a *water rail* or *common snipe*, especially in the autumn and winter. Look out too for the *grey heron*, and of course for many of the migrating birds that are just passing through, such as *swallows* and *martins. Kestrels* and the *curlew* are often seen in the more open areas and in winter *fieldfares* and *redwings* can usually be seen, together with *stonechats*, which like to perch on the overhead wires.

I have concentrated on the Lower Moors Nature Trail because it is such a delightful place to visit while on St Mary's, and because it encompasses so much of the natural history of the island. Before leaving it, do see if you can spot the *eels* and *grey mullet* in the stream that flows south from the main pool to the sea at Old Town. At high tide, sea water enters the main ditch up as far as the dam just south of the bird hides, and it is in this stretch that you are more likely to see them in the clear water. Other nature reserves worth exploring on St Mary's are the wetlands at Higher Moors and Holy Vale.

Two uncommon plants are to be found on St Mary's in quite large numbers, namely *western fumitory* and *sea radish*. Look for the somewhat floppy and straggly plant of western fumitory, about 20 to 30 centimetres high, amongst the bulb fields and on walls from May to September. The pinkish-purple flowers are made up of four petals which are dark purple at the tips and are joined together to form a tube about a centimetre long. This species is confined on the mainland, chiefly to Cornwall, and was first recorded in Scilly in 1937. There are some 15 native species of fumitory in Britain, of which about 5 are found on the islands. The *common fumitory*, which is smaller and less handsome than the western fumitory, is an abundant weed in the cultivated fields of the inhabited islands. Its smoke-grey feathery leaves, deeply divided into narrow segments, are the origin of its medieval name *fumus terrae*, meaning smoke of the earth. When the plant is pulled up it smells of smoke, and its juice makes the eyes water; in ancient times it had a reputation as a cure for all ills.

Sea radish is a straggly plant some 120 centimetres tall with pale yellow flowers that forms colonies along the drift lines of sandy and rocky shores. Its lower leaves are large, lobed and toothed and the whole plant is covered with rough hairs; it flowers throughout the summer. At Porth Cressa you may find the white flowering variety of this fairly local plant, which is confined to St Mary's and St Agnes.

A plant that has established itself on rocks and walls on many of the inhabited islands is the *wall oxalis*. A native of Chile, where it is common on coastal rocks, it was introduced to Tresco Abbey Gardens in 1879 and has spread and become naturalised. It has thick, fleshy stems and attractive bright yellow flowers, and its bright green oval leaves are formed in threes in typical oxalis style. Look for wall oxalis on the walls around Hugh Town and the rocks about Porth Cressa. In late April around Pelistry Bay, don't forget to look out for the attractive violet-blue flowers of the *spring squill*, as described in Chapter 7.

A very rare plant growing only in one or two places on St Mary's is the *common fleabane*. This is often seen on the mainland in marshy places and it is surprising that it has not spread more widely in Scilly. It springs from a creeping rootstock and has branching woolly stems about 30 centimetres tall; the oblong heart-shaped leaves are wrinkled and downy and are arranged alternately down the stem. The deep yellow flower heads, about 3 centimetres across, are produced at the end of the branches and bloom from July to October. When burnt, the smoke of the plant was said to drive away fleas and gnats, hence its English name. Formerly this plant was used as a medicine for dysentery and this is the reason for its Latin name, *dysenterica*. Look for this attractive plant in the damp areas on Higher Moors, and perhaps elsewhere.

Three butterflies are particularly attracted to the common fleabane, namely the *small tortoiseshell*, the *small copper* (both already described in Chapters 2 and 10 respectively) and the *small heath*.

With a wingspan of just less than 3 centimetres, the small heath is a light brown butterfly which you will see along the roadside verges and ditches between May and September, more commonly in the later months. It has faint dark spots at the tips of the forewings and on the underside there is a prominent false eye on each of the forewings. The underside of the hind wings is darker near the base and each has a row of faint circles. The butterflies live for about a month, and the warm climate of the south of England and the Isles of Scilly allows for two generations a year. The first eggs are laid singly in grasses in May and these hatch out into caterpillars two weeks later. The caterpillar stage can last just a month or it can develop over a period as long as 11 months, which includes a period of hibernation in the winter. The caterpillars feed on grasses and those hatching out in May produce butterflies that emerge from the chrysalids in August. Caterpillars hibernating and feeding on grass in mild weather during winter produce butterflies the following May.

So we leave the butterflies of the sun-washed open spaces and bright wayside flowers and turn to one that is more common on the shady paths of woodland glades. In such places in early spring look for the *speckled wood* butterfly, its brown and buff speckled wings giving it camouflage in the dappled sunlight that percolates the leaf canopy of its habitat. The male, which has smaller buff spots than the female, settles in patches of sunlight where a female is more likely to see it and be attracted by it. The speckled wood shows a high degree of territorial behaviour, and the male will defend its patch of woodland by attacking any intruder. Sit here in this secluded spot amongst the brambles and bushes near the narrow winding lane and watch a male fly in a series of up and down movements to harry another that dares to intrude on its territory. The two butterflies tussle together, spiralling upwards in their mutual anger, but it is the defender that is usually successful in seeing off the other. As with most butterflies, scent produced from the base of the wing scales plays an important part in the speckled wood's short lifespan of only 20 days. The male almost certainly marks its territory by scent, which is also a possible attraction to the female. The eggs are laid twice a year at any time from spring to autumn on the leaves of various grasses; the small green caterpillars emerge in 10 days and then feed at night on the grasses for about 30 days. They then turn into chrysalids, before the new butterfly emerges a month later. There are usually two generations a year in Britain, and autumn caterpillars live throughout the winter, actively feeding on warm days. Interestingly, like the *meadow brown* and *Tean blue* mentioned in Chapter 4, the speckled wood on The Isles of Scilly is a distinct subspecies of the butterfly, with its own distinct markings on its wings. Such is the world of butterflies that even within the subspecies there are variations depending on when the adults emerge from their chrysalids. Those of the speckled wood that fly in late March or April have unusually large white markings, whereas those that emerge in May or in late summer have smaller markings. Clearly, temperature and genetics play their part in this seasonal variation, which is not restricted solely to the speckled wood but is common in other species such as the *comma* and *white* butterflies. In the case of the speckled wood, perhaps the absence of a good leaf canopy in the early spring means that more sunlight is apparent in the butterflies' habitat, and hence the larger the white patches on its speckled wings, the better the camouflage; the reverse being true as the season progresses.

Many other of the 54 or so known British butterflies will be seen flitting amongst the byways and waste places of this large island. Look out for the comma butterfly, basking for long periods with its black-spotted orange-brown wings spread wide, soaking up the sunshine. Both sexes have a prominent white 'comma' mark on their undersides and it is the only British butterfly that has ragged edges to its wings; a design that has evolved through natural selection to provide camouflage amongst the brambles, thistles and other vegetation which it favours. Two generations are usual, the first on the wing in July are lighter in colour than the second, which fly in September and October.

Of the white species, see if you can recognise the *green-veined white* with its distinctive dark green veins under its creamy-white wings, which stand out more heavily in the summer than in the spring. Even on cloudy days you may see this butterfly flying over damp meadows, marshy land and wayside ditches near hedgerows. The male of the species, distinguished by having usually only one black spot on its upper forewing compared to the two spots of the female, exudes a strong scent of lemon, probably used in courtship.

From the world of butterflies we turn briefly to animals and specifically to that much studied mammal, the *lesser white-toothed shrew*, designated appropriately in Latin *Crocidura cassiteridum*. Unknown elsewhere in Britain but present in considerable numbers on the inhabited islands of Scilly, this species can be distinguished from all other British shrews by its white teeth and greyish or silvery-brown fur. They were discovered in 1924 and it was thought at first that they might have migrated from the Channel Islands or Brittany (where there is a similar but different species) many thousands of years ago, when there was still a land bridge. However, a more likely explanation is that they came over with early man when he first inhabited the islands soon after the ice sheets melted. Very few animals would have been needed to start the colonisation of Scilly, even one pregnant female would have sufficed, and a population once established would inevitably develop some differences from the mainland population due to the limited gene bank of the founder stock. Whatever the reason, we have on Scilly a distinct species which inhabits gardens, farmland and wooded areas.

Shrews feed on a wide range of small animals and insects such as earthworms, beetles, caterpillars, spiders, flies and woodlice; their saliva is toxic, and when they bite with their sharp teeth, the prey is dead within minutes. They mate between the early spring and summer and up to five litters of seven or eight young can be born during that period. The nest of the shrew is a ball of woven grass placed at the end of a tunnel burrowed into the ground in undergrowth or leaf litter. The young are suckled by their mother for the first three or four weeks and are then left to fend for themselves.

Other animals you could see on St Mary's are the *slow-worm*, a member of the lizard family, and the *frog*. The slow-worm is not common and was probably introduced, like the hedgehog, but the frog is probably indigenous; strangely there are no toads on Scilly. Frogs are plentiful and live just about anywhere near water, in damp grass and undergrowth. These amphibians, which in the winter hibernate under bushes, in deep grass and under stones or logs, make their way on the first warm day of late winter or early spring to damp meadows, ditches and pools. It is at this time that the males find their voices, their throats swell and the wet places become alive with their croaking. Females respond with chirps and grunts, and their thousands of eggs are laid in the water in the form of jelly-like spawn. Frog's eggs hatch in May into tadpoles and by the end of June they have changed into tiny 2 centimetre miniature adults. They then jump out of the water and disappear into the surrounding vegetation, feeding on small insects, slugs, snails and earthworms. Within three years they are fully grown adults able to mate and reproduce. Frogs form an important part of the food chain as their eggs, tadpoles and adults are preyed upon by fish, rats and hedgehogs, and particularly by birds such as the heron and little egret.

It is time to say goodbye to St Mary's and indeed to all The Isles of Scilly. Take one final walk to Porth Hellick Point for a moment of quiet contemplation amongst the ancient entrance graves and cairns of long-forgotten people; below on top of the beach, the small monument to Sir Cloudesley Shovell reminds you of the less distant past and of that great British fleet wrecked on the Western Rocks. A group of *oystercatchers* catch your sense of melancholy with their wild plaintive piping as they look forlornly out to sea waiting for low tide. The sea ever pounding on the rocks below emphasises the permanence of this scene that will certainly last for another thousand years. With great reluctance you make your way to the airport, with its bustle of humanity and cars, its café and small tables, and people seemingly imprisoned until their flight is called, when they make their next move back to the stress of mainland living. You follow with hesitancy, unwilling to leave behind the islands that gave you so much pleasure in the past ten days. The flight takes off with the sun sinking gently to a silvery sea. Farewell Tresco and Bryher, with your pools and downs where peace and beauty are all around; farewell too, you far-flung Western Rocks, where only seals and seabirds live amongst those surging white lines of spray and enjoy their surrounding ambience. Goodbye Great Arthur and Samson, with your history and spirits of ancient peoples.

> Cries of lonely single gulls
> Soar through waves of salty air;
> Glistening ripples of water lull
> Basking in a golden flair, as
> The serene ball of fire sets
> Shimmering over the velvet hills
> Sending great beauty to the isles.
> The peace – so tranquil.

Jane Coulcher
1998

EPILOGUE

In our journey through the beautiful Isles of Scilly I am very conscious that I have just gently touched the face of nature and inadequately described its friendly people and their history; there is so much more to see and hear. Throughout, I have tried to keep the reader moving on and to link together the islands with their beauty and their separate individuality and character. From the tranquillity of Tresco's Abbey Pool and through the flower-strewn bulb fields of St Martin's early on a sunny June morning, to the dreadful impetuosity of pounding waves on the Western Rocks, I hope I have kept your interest in what is a remarkable and unique place on this planet. I hope that if people read this book in 50 or 100 years time, they will be able to say, 'It hasn't changed much since he wrote about it in 1998.' But this will depend on constant vigilance by bodies such as The Duchy of Cornwall, The Council and The Isles of Scilly Environmental Trust to prevent such disharmony as noise from powerboats and water-skiing, pursuits that have blighted many otherwise tranquil places on the mainland. The Islands are still a haven for an unequalled kind of peace and restfulness that has almost vanished on mainland coasts, so let them continue to reflect a last slow-paced age, unique in these modern times.

Of course, you the reader will not see everything described in this book – that would be almost impossible – but you will see many of the birds and flowers, the grandeur of the rocks and cliffs and always the blue sea and sky. If you are lucky to view the cavortings of the raven, seal pups on low sea ledges and perhaps a basking shark, then that would be a big bonus. Just take pleasure in it all, and even if the soft mist and rain descends, remember that contrast is essential to the full enjoyment of life.

How does one conclude a book such as this; with a happy contentment that one has tried to give pleasure to the reader, or with sadness that the journey is at an end and there is no more to relate? My conclusion will touch on the sea, that great ocean that surrounds the Islands and that has always had such a profound effect on its people and their way of life. Remember the times you wandered under blue skies and then sat down on one of the Islands' fairest beaches and gazed out to a horizon far away from that sunlit island shore; remember too the rhythm of the tides that lulled your senses and recall some lines from John Masefield's great poem 'Sea Fever':

I must go down to the seas again, for the call of the running tide,
Is a wild call and a clear call that may not be denied;
And all I ask is a windy day with the white clouds flying,
And the flung spray and the blown spume, and the sea-gulls crying.

So with a great peace tinged with sadness we leave these Sun Islands, these glorious Isles of Scilly, where there still can be found a deep silence, empty of all but the wind and sea.

Sunset over Guther's Island

BIBLIOGRAPHY

AA Book of the Countryside, Drive Publications 1973

Birds, Trees and Flowers, Odhams Press 1947

Bowley, R.L., *The Fortunate Islands*, Bowley Publications 1990

Brown, Leslie, *British Birds of Prey*, Collins 1976

Cornwall Archaeological Unit, *Scilly Archaeological Heritage*, Twelveheads Press 1995

Coulcher, Patrick, *A Natural History of the Cuckmere Valley*, The Book Guild 1998

Coulcher, Patrick, *The Mountain of Mist*, The Book Guild 1998

Fisher, James, and Lockley RM, *Sea Birds*, Collins 1954

Fisher, James, *The Fulmar*, Collins 1952

Ford, E.B. *Butterflies*, Collins 1945

Gibson, Frank, *Sea and Shore Birds and Marine Life on the Shores of the Isles of Scilly*, Beric Tempest

Gibson, Frank, *Visitors' Companion to the Isles of Scilly*, Beric Tempest

*Gibson, Frank, *Wild Flowers of Scilly*, Beric Tempest

Hepburn, Ian, *Flowers of the Coast*, Collins 1952

Hewer, H.R., *British Seals*, Collins 1954

Imms, A.D., *Insect Natural History*, Collins 1947

Isles of Scilly Environmental Trust, *A Precious Heritage*, Beric Tempest

Kearton, Richard, *British Birds' Nests*, Cassell 1908

Lousley, J.E., *The Flora of the Isles of Scilly*, David Charles 1971

Lousley, J.E., *Flowering Plants and Ferns in the Isles of Scilly*, Isles of Scilly Museum Publication no 4 1983

Matthews, L. Harrison, *British Mammals*, Collins 1952

A NATURAL HISTORY OF THE ISLES OF SCILLY

*Martin, W. Keble, *The Concise British Flora in Colour*, Rainbird 1965

Mumford, Clive, *Portrait of the Isles of Scilly*, Roberthale and Co. 1972

Over, Luke, *The Kelp Industry in the Isles of Scilly*, Isles of Scilly Museum Publication No. 14, 1987

*Peterson, Roger, Mountfort, Guy and Hollom, P.A.D., *A Field Guide to the Birds of Britain and Europe*, Collins 1954

Pitt Frances, *Birds of Britain*, Macmillan 1948

*Reader's Digest, *Field Guide to the Butterflies and Other Insects of Britain*, 1984

Robinson, Peter, *Birds in the Isles of Scilly*, Isles of Scilly Museum Publication No. 2 1993

Step, Edward, *Wayside and Woodland Blossoms*, Frederick Warne 1941

Wickham, Cynthia, *Common Plant as Natural Remedies*, Frederic Muller 1981

*These are useful books for identification purposes.

BIRDS IN THE ISLES OF SCILLY – A CHECK LIST

Extract from Publication No. 2, revised by Peter Robinson, 1993.
By kind permission of the Isles of Scilly Museum

DIVERS: Gaviidae
RED-THROATED DIVER Gavia Stellata – Scarce migrant
BLACK-THROATED DIVER Gavia arctica – Scarce migrant or winter visitor
GREAT NORTHERN DIVER Gavia immer – Regular winter visitor in small numbers
WHITE-BILLED DIVER Gavia adamsii – Extremely rare vagrant

GREBES: PODICIPEDIDAE
LITTLE GREBE Tachybaptus ruficollis – A few visit fresh water pools in winter
GREAT-CRESTED GREBE Podiceps cristatus – Very rare vagrant
RED-NECKED GREBE Podiceps griseigena – Very rare vagrant
SLAVONIAN GREBE Podiceps auritus – Very rare vagrant
BLACK-NECKED GREBE Podiceps nigricollis – Scarce migrant or winter visitor

ALBATROSSES: Diomedeidae
BLACK-BROWED ALBATROSS Diomedea
melanophris – Extremely rare vagrant S.Atl

FULMAR, SHEARWATERS: Procellariidae
FULMAR Fulmarus glacialis – Breeding summer visitor in small numbers
CORY'S SHEARWATER Puffinus diomedea – Rare offshore migrant Med or S.Atl
GREAT SHEARWATER Puffinus gravis – Rare offshore migrant S.Atl
SOOTY SHEARWATER Puffinus grisea – Rare offshore migrant S.Atl or Pac
MANX SHEARWATER Puffinus puffinus – Breeding summer visitor
BALEARIC SHEARWATER Puffinus puffinus mauretanicus (sub-species) – Rare offshore vagrant Med
LITTLE SHEARWATER Puffinus assimilis – Extremely rare offshore vagrant

PETRELS: Hydrobatidae
STORM PETREL Hydrobates pelagicus – Breeding summer visitor
LEACH'S PETREL Oceanodroma leucorhoa – Rare offshore migrant

GANNETS: Sulidae
GANNET Sula bassana – All year round at sea

CORMORANTS, SHAGS: Phalacrocoracidae
CORMORANT Phalacrocorax carbo – Common breeding resident in small numbers

SHAG Phalacrocorax aristotelis – Common breeding resident

HERONS, BITTERNS, EGRETS: Ardeidae
BITTERN Botaurus stellaris – Very rare vagrant
AMERICAN BITTERN Botaurus lentiginosus – Extremely rare vagrant N.Am
LITTLE BITTERN Ixobrychus minutus – Very rare vagrant
NIGHT HERON Nycticorax nycticorax – Rare vagrant S.Eur
SQUACCO HERON Ardeaola ralloides – Very rare vagrant S.Eur
LITTLE EGRET Egretta garzetta – Rare vagrant S.Eur. commoner in recent times
GREY HERON Ardea cinerea – Regular migrant and winter visitor
PURPLE HERON Ardea purpurea – Rare migrant S.Eur

STORKS: Ciconiidae
BLACK STORK Ciconia nigra – Very rare vagrant
WHITE STORK Ciconia ciconia – Very rare vagrant

IBISES, SPOONBILLS: Threskiornithidae
GLOSSY IBIS Plegadis falcinellus – Very rare vagrant
SPOONBILL Platalea leucorodia – Rare vagrant

SWANS, GEESE, DUCKS: Anatidae
MUTE SWAN Cygnus olor – Scarce breeding resident
BEWICK'S SWAN Cygnus columbainus – Rare winter vagrant
WHOOPER SWAN Cygnus cygnus – Rare winter visitor
BEAN GOOSE Anser fabalis – Very rare vagrant
PINK-FOOTED GOOSE Anser brachyrhynchus – Rare migrant or winter visitor
WHITE-FRONTED GOOSE Anser albifrons – Rare migrant or winter visitor
GREYLAG GOOSE Anser anser – Rare vagrant
CANADA GOOSE Branta canadensis – Scarce breeding resident (introduced 1920s)
BARNACLE GOOSE Branta leucopsis – Rare migrant or winter visitor
BRENT GOOSE Branta bernicla – Rare vagrant or winter visitor
SHELDUCK Tadorna tadorna – Scarce breeding summer visitor, winter visitor and migrant
WIDGEON Anas penelope – Winter visitor
AMERICAN WIDGEON Anas americana – Very rare vagrant N.Am
GADWALL Anas strepera – Breeding resident since 1934
TEAL Anas crecca – Winter visitor, some also summer and occasionally breed
GREEN-WINGED TEAL Anas crecca carolinensis (sub-species) – Very rare vagrant N.Am
MALLARD Anas platyrhynchos – Breeding resident and winter visitor
BLACK DUCK Anas rubripes – Extremely rare vagrant N.Am
PINTAIL Anas acuta – Occasional migrant or winter visitor
GARGANY Anas querquedula – Scarce migrant, has bred
BLUE-WINGED TEAL Anas discors – Very rare vagrant N.Am
SHOVELLER Anas clypeata – Scarce resident and winter visitor
RED-CRESTED POCHARD Netta rufina – Extremely rare vagrant
POCHARD Aythya ferina – Winter visitor and migrant
RING-NECKED DUCK Aythya collaris – Very rare vagrant N.Am
FERRUGINOUS DUCK Aythya nyroca – Extremely rare vagrant
TUFTED DUCK Aythya rufigula – Winter visitor, migrant and scarce breeder
SCAUP Aythya marila – Rare winter visitor and migrant

BIRDS IN THE ISLES OF SCILLY

EIDER Somateria mollissima – Rare winter migrant
LONG-TAILED DUCK Clangula hyemalis – Scarce winter visitor
COMMON SCOTER Melanitta nigra – Scarce winter visitor and migrant
SURF SCOTER Malanitta perspecillata – Very rare vagrant N.Am
VELVET SCOTER Melanitta fusca – Very rare vagrant
BUFFELHEAD Bucephala albeola – Extremely rare vagrant N.Am
GOLDENEYE Bucephala clangula – Scarce winter visitor and migrant
SMEW Mergus albellus – Very rare vagrant
RED-BREASTED MERGANSER Mergus serrator – Scarce winter visitor and migrant
GOOSANDER Mergus merganser – Very rare vagrant
RUDDY DUCK Oxyura jamaicensis – Extremely rare vagrant

OSPREY: Pandionidae
OSPREY Pandion haliaetus – Rare migrant

EAGLES, BUZZARDS, HAWKS, KITES, HARRIERS, FALCONS: Falaconidae
HONEY BUZZARD Pernis apivorus – Rare migrant
BLACK KITE Mivus migrans – Very rare vagrant
RED KITE Milvus milvus – Rare vagrant or migrant
WHITE-TAILED EAGLE Haliaeetus albicilla – Extremely rare vagrant
MARSH HARRIER Circus aeruginosus – Scarce migrant
HEN HARRIER Circus cyaneus – Scarce migrant
MONTAGU'S HARRIER Circus pygargus – Rare migrant
GOSHAWK Accipiter gentilis – Very rare vagrant
AMERICAN GOSHAWK Accipiter gentilis atricapillus (sub-species) – Extremely rare vagrant N.Am
SPARROWHAWK Accipiter nisus – Scarce winter visitor and migrant
BUZZARD Buteo buteo – Rare migrant
ROUGH-LEGGED BUZZARD Buteo lagopus – Very rare vagrant
LESSER KESTREL Falco naumanni – Very rare vagrant
KESTREL Falco tinnunculus – Scarce breeding resident, increases in winter
RED-FOOTED FALCON Falco vespertinus – Very rare vagrant
MERLIN Falco columbarius – Regular winter visitor in small numbers
HOBBY Falco subbuteo – Scarce migrant
GYR FALCON Falco rusticolus – Very rare vagrant
PEREGRINE FALCON Falco peregrinus – Scarce migrant and former breeder

QUAILS, PHEASANTS: Phasianidae
QUAIL Coturnix coturnix – Irregular migrant in small numbers, has bred
[BOB-WHITE QUAIL Colinus virginianus] – Introduced to Tresco but extinct there by late 1970s [N.Am]
PHEASANT Phasianus colchicus – Introduced breeder but remaining scarce away from Tresco
GOLDEN PHEASANT Chrysolophus pictus – Introduced breeder Tresco 1970s, remains scarce [China]

CRAKES, RAILS, GALLINULES: Rallidae
WATER RAIL Rallus aquaticus – Migrant and winter visitor, has bred
SPOTTED CRAKE Porzana porzana – Scarce migrant

SORA RAIL Porzana carolia – Very rare vagrant N.Am
LITTLE CRAKE Porzana parva – Extremely rare vagrant
CORNCRAKE Crex crex – Scarce migrant and former breeder
MOORHEN Gallinula chloropus – Breeding resident
AMERICAN PURPLE CALLINULE Porphyrula martinica – Extremely rare vagrant N.Am
COOT Fulica atra – Breeding resident and migrant, increases in winter

CRANES: Gruidae
COMMON CRANE Grus grus – Extremely rare vagrant

BUSTARDS: Otididae
LITTLE BUSTARD Tetrax tetrax – Extremely rare vagrant

OYSTERCATCHERS: Haematopodidae
OYSTERCATCHER Haematopus ostralegus – Common breeding resident and migrant

STILTS, AVOCETS: Recurvirostridae
BLACK-WINGED STILT Himantopus himantopus – Extremely rare vagrant
AVOCET Recurvirostra avocetta – Very rare vagrant

STONE CURLEWS: Burhinidae
STONE CURLEW Burhinus oedicnemus – Very rare vagrant

PLOVERS: Charadriidae
LITTLE RINGED PLOVER Charadrius dubius – Regular scarce migrant
RINGED PLOVER Charadrius hiaticula – Common breeding resident, migrant and winter visitor
SEMIPALMATED PLOVER Charadrius semipalmatus – Extremely rare vagrant N.Am
KILLDEER Charadrius vociferus – Very rare vagrant N.Am
KENTISH PLOVER Charadrius alexandrinus – Extremely rare vagrant or migrant
CASPIAN PLOVER Charadrius asiaticuas – Extremely rare vagrant C.As
DOTTEREL Charadrius morinellus – Regular scarce migrant
AMERICAN GOLDEN PLOVER Pluvialis dominica – Rare vagrant N.Am
GOLDEN PLOVER Pluvialis apricaria – Migrant and winter visitor
GREY PLOVER Pluvialis squatarola – Migrant and winter visitor
LAPWING Vanellus vanellus – Migrant and winter visitor

SNIPES, CURLEWS, GODWITS, STINTS, SANDPIPERS, DOWITCHERS, ETC:
Scolopacidae
KNOT Calidris canutus – Regular scarce migrant
SANDERLING Calidris alba – Migrant and winter visitor
SEMI-PALMATED SANDPIPER Calidris pusilla – Very rare vagrant N.Am
LITTLE STINT Calidris minuta – Scarce migrant
TEMMINCK'S STINT Calidris temminckii – Very rare vagrant or migrant
LEAST SANDPIPER Calidris minutilla – Extremely rare vagrant N.Am
WHITE-RUMPED SANDPIPER Calidris fuscicollis – Rare vagrant N.Am
BAIRD'S SANDPIPER Calidris bairdii – Very rare vagrant N.Am
PECTORAL SANDPIPER Calidris melanotos – Regular scarce migrant or vagrant N.Am

SHARP-TAILED SANDPIPER Calidris acuminata – Extremely rare vagrant N.Am
CURLEW SANDPIPER Calidris ferruginea – Scarce regular migrant
PURPLE SANDPIPER·Calidris maritima – Winter visitor and migrant
DUNLIN Calidris alpina – Winter visitor and migrant
BUFF-BREASTED SANDPIPER Tryngites subruficollis – Scarce but almost annual vagrant N.Am
RUFF Philomachus pugnax – Scarce migrant
JACK SNIPE Lymnocryptes minimus – Migrant and scarce winter visitor
SNIPE Gallinago gallinago – Migrant and winter visitor
GREAT SNIPE Gallinago media – Extremely rare vagrant
LONG-BILLED DOWITCHER Limnodromus scolopaceus – Very rare vagrant N.Am
WOODCOCK Scolopax rusticola – Winter visitor and migrant
BLACK-TAILED GODWIT Limosa limosa – Scarce migrant
BAR-TAILED GODWIT Limosa lapponica – Winter visitor and migrant
ESKIMO CURLEW Numenius borealis – One record, shot Tresco 1887 (last European record) Possibly extinct
WHIMBREL Numenius phaeopus – Migrant
CURLEW Numenius arquata – Winter visitor and migrant, a few summer
UPLAND SANDPIPER Bartramia longicauda – Very rare vagrant N.Am
SPOTTED REDSHANK Tringa erythropus – Scarce migrant
REDSHANK Tringa totanus – Winter visitor and migrant
GREENSHANK Tringa nebularia – Winter visitor and migrant
GREATER YELLOWLEGS Tringa melanoleuca – Very rare vagrant N.Am
LESSER YELLOWLEGS Tringa flavipes – Very rare vagrant N.Am
SOLITARY SANDPIPER Tringa solitaria – Very rare vagrant N.Am
GREEN SANDPIPER Tringa ochropus – Migrant
WOOD SANDPIPER Tringa glareola – Scarce migrant
COMMON SANDPIPER Actitis hypoleucos – Migrant
SPOTTED SANDPIPER Actitis macularia – Rare vagrant
TURNSTONE Arenaria interpres – Winter visitor and migrant

PHALAROPES: Phalaropodidae
WILSON'S PHALAROPE Phalaropus tricolor – Very rare vagrant N.Am
RED-NECKED PHALAROPE Phalaropus lobatus – Rare migrant
GREY PHALAROPE Phalaropus fulicarius – Winter visitor and migrant

PRATINCOLES: Glareolidae
COLLARED PRATINCOLE Glareola pratincola – Extremely rare vagrant

SKUAS: Stercorariidae
POMARINE SKUA Stercorarius pomarinus – Scarce offshore migrant
ARCTIC SKUA Stercorarius parasiticus – Scarce migrant offshore and between the islands
LONG-TAILED SKUA Stercorarius longicaudus – Rare offshore migrant
GREAT SKUA Stercorarius skua – Regular offshore migrant

GULLS, KITTIWAKES, TERNS: Laridae
MEDITERRANEAN GULL Larus melanocephalus – Rare vagrant
LAUGHING GULL Larus aticilla – Extemely rare vagrant N.Am
LITTLE GULL Larus minutus – Scarce migrant
SABINE'S GULL Larus sabini – Rare migrant or vagrant N.Am

BONAPARTE'S GULL Larus philadelphia – Very rare vagrant N.Am
BLACK-HEADED GULL Larus ridibundus – Migrant and winter visitor
RING-BILLED GULL Larus delawarensis – Very rare vagrant N.Am
COMMON GULL Larus canus – Migrant and winter visitor
LESSER BLACK-BACKED GULL Larus fuscus – Common breeding summer visitor
HERRING GULL Larus argentatus – Common breeding resident and partial migrant
ICELAND GULL Larus glaucoides – Rare vagrant
GLAUCOUS GULL Larus hyperboreus – Rare vagrant
GREAT BLACK-BACKED GULL Larus marinus – Common breeding resident and partial migrant
KITTIWAKE Rissa tridactyla – Breeding summer visitor and migrant
IVORY GULL Pagophilia eburnea – Extremely rare vagrant
GULL-BILLED TERN Gelochelidon nilotica – Extremely rare vagrant
SANDWICH TERN Sterna sandvicensis – Migrant and scarce breeding summer visitor
ROSEATE TERN Sterna dougallii – Migrant and scarce breeding summer visitor
COMMON TERN Sterna hirondo – Migrant and breeding summer visitor
ARCTIC TERN Sterna paradisaea – Uncommon migrant, has bred
BRIDLED TERN Sterna anaethetus – Extremely rare vagrant. Red Sea, Caribbean, West Africa or Indian Ocean
LITTLE TERN Sterna albifrons – Scarce migrant
WHISKERED TERN Chlidonias hybridus – Extremely rare migrant
BLACK TERN Chlidonias niger – Scarce migrant
WHITE-WINGED BLACK TERN Chlidonias leucopterus – Extremely rare migrant

AUKS: Alcidae
GUILLEMOT Uria aalge – Breeding summer visitor, a few winter
RAZORBILL Alca torda – Breeding summer visitor, a few winter
BLACK GUILLEMOT Cepphus grylle – Extremely rare vagrant
LITTLE AUK Alle alle – Scarce winter visitor
PUFFIN Fratercula arctica – Breeding summer visitor

SANDGROUSE: Pteroclididae
PALLAS'S SANDGROUSE Syrrhaptes paradoxus – Extremely rare vagrant C.As

PIGEONS, DOVES: Columbidae
FERAL PIGEON Columba livia – Resident
STOCK DOVE Columba oenas – Scarce resident breeder, migrant and winter visitor
WOOD PIGEON Columba palumbus – Breeding resident, winter visitor and migrant
COLLARED DOVE Streptopelia decaocto – Common breeding resident and partial migrant
TURTLE DOVE Streptopelia turtur – Migrant and occasional breeder
RUFOUS TURTLE DOVE Streptopelia orientalis – Extremely rare vagrant C.As

PARROTS, PARAKEETS ETC: Psittaculadae
RING-NECKED PARAKEET Psittacula kramerei – Extremely rare vagrant; either from feral mainland population or as escape

CUCKOOS: Cuculidae
GREAT SPOTTED CUCKOO Clamator glandarius – Extremely rare vagrant S.Eur
CUCKOO Cuculus canorus – Breeding summer visitor and migrant

BIRDS IN THE ISLES OF SCILLY

BLACK-BILLED CUCKOO Coccyzus erythropthalmus – Extremely rare vagrant N.Am
YELLOW-BILLED CUCKOO Coccyzus americanus – Very rare vagrant N.Am

OWLS: Strigidae
BARN OWL Tyto alba – Rare winter visitor
SCOPS OWL Otus scops – Extremely rare vagrant S.Eur
SNOWY OWL Nyctea scandiaca – Very rare vagrant
LITTLE OWL Athene noctua – Extemely rare vagrant
TAWNY OWL Strix aluco – Very rare vagrant
LONG-EARED OWL Asio otus – Rare migrant
SHORT-EARED OWL Asio flammeus – Scarce migrant

NIGHTJARS: Caprimulgidae
NIGHTJAR Caprimulgus europaeus – Scarce migrant and former breeder
COMMON NIGHTHAWK Chordeiles minor – Very rare vagrant N.Am

SWIFTS: Apodidae
CHIMNEY SWIFT Chaetura pelagica – Extremely rare vagrant N.Am
SWIFT Apus apus – Migrant and summer visitor, has bred
ALPINE SWIFT Apus melba – Rare vagrant

KINGFISHERS: Alcedinidae
KINGFISHER Alcedo atthis – Scarce migrant, occasionally winters

BEE-EATERS: Meropidae
BLUE-CHEEKED BEE-EATER Merops superciliosus – Extremely rare vagrant
BEE-EATER Merops apiaster – Rare vagrant

ROLLERS: Coraciidae
ROLLER Coracias garrulus – Very rare vagrant S.Eur

HOOPOES: Upupidae
HOOPOE Upupa epops – Scarce migrant

WOODPECKERS: Picidae
WRYNECK Jynx torquilla – Scarce migrant
GREEN WOODPECKER Picus viridis – Extremely rare vagrant
YELLOW-BELLIED SAPSUCKER Sphyrapicus varius – Extremely rare vagrant
GREAT SPOTTED WOODPECKER Dendrocopos major – Very rare vagrant

LARKS: Alaudidae
CALANDRA LARK Melanocorypha calandra – Extremely rare vagrant
BIMACULATED LARK Melanocorypha bimaculata – Extremely rare vagrant
SHORT-TOED LARK Calandrella cinerea – Rare vagrant
WOODLARK Lullula arborea – Rare migrant
SKYLARK Alauda arvensis – Scarce breeding summer visitor, migrant and winter visitor
SHORE LARK Eremophilia alpestris – Very rare vagrant

157

A NATURAL HISTORY OF THE ISLES OF SCILLY

SWALLOWS, MARTINS: Hirundinidae
SAND MARTIN Riparia riparia – Migrant, has bred
TREE SWALLOW Tachycineta bicolor – Extremely rare vagrant N.Am
SWALLOW Hirundo rustica – Breeding summer visitor and common migrant
RED-RUMPED SWALLOW Hirundo daurica – Rare vagrant
CLIFF SWALLOW Hirundo pyrrhonota – Extremely rare vagrant N.Am
HOUSE MARTIN Delichon urbica – Scarce breeding summer visitor and common migrant

PIPITS, WAGTAILS: Motacillidae
RICHARD'S PIPIT Anthus novaeseelandiae – Scarce annual vagrant
TAWNY PIPIT Anthus campestris – Scarce annual vagrant
OLIVE-BACKED PIPIT Anthus hodgsoni – Rare vagrant
TREE PIPIT Anthus trivialis – Migrant
MEADOW PIPIT Anthus pratensis – Migrant and winter visitor, occasional breeder
RED-THROATED PIPIT Anthus cervinus – Rare vagrant
ROCK PIPIT Anthus petrosus – Common breeding resident and migrant
WATER PIPIT Anthus spinoletta – Rare migrant
BUFF-BELLIED PIPIT Anthus rubescens – Extremely rare vagrant N.Am
BLUE-HEADED WAGTAIL Motacilla flava – Scarce annual migrant
YELLOW WAGTAIL Motacilla flava flavissima – Migrant and the commonest of this 'yellow' wagtail group
GREY-HEADED WAGTAIL Motacilla flava thunbergi (sub-species) – Very rare migrant
BLACK-HEADED WAGTAIL Motacilla flava feldegg (sub-species) – Extremely rare vagrant
CITRINE WAGTAIL Motacilla citreola – Very rare vagrant
GREY WAGTAIL Motacilla cinerea – Migrant, occasionally winters
WHITE WAGTAIL Motacilla alba – Migrant and winter visitor, had bred
PIED WAGTAIL Motacilla alba yarrelli (sub-species) – Migrant

WAXWINGS: Bombycillidae
WAXWING Bombycilla garrulus – Rare migrant or winter visitor

WRENS: Troglodytidae
WREN Troglodytes troglodytes – Common and widespread breeding resident

ACCENTORS: Prunellidae
DUNNOCK Prunella modularis – Common breeding resident
ALPINE ACCENTOR Prunella collaris – Extremely rare vagrant

THRUSHES, WHEATEARS, CHATS: Turdidae
ROBIN Erithacus rubecula – Common breeding resident and migrant
THRUSH NIGHTINGALE Luscinia luscinia – Extremely rare migrant
NIGHTINGALE Luscinia megarhynchos – Scarce migrant
BLUETHROAT Luscinia svecica – Scarce migrant
BLACK REDSTART Phoenicurus ochruros – Migrant and winter visitor
REDSTART Phoenicurus phoenicurus – Migrant

BIRDS IN THE ISLES OF SCILLY

WHINCHAT Saxicola rubetra – Migrant
STONECHAT Saxicola torquata – Breeding resident and migrant
STONECHAT Saxicola torquata stejnegeri/maura (sub-species) – Rare vagrant
ISABELLINE WHEATEAR Oenanthe isabelline – Very rare vagrant C.As or M.E
NORTHERN WHEATEAR Oenanthe oenanthe – Migrant and scarce breeder
BLACK-EARED WHEATEAR Oenanthe hispanica – Very rare vagrant S.Eur or M.E
DESERT WHEATEAR Oenanthe deserti – Extremely rare vagrant C.As or M.E
ROCK THRUSH Monticola saxatilis – Very rare vagrant S.Eur
[BLUE ROCK THRUSH Monticola solitarius] – Not accepted to the British List; one on St Martin's was thought an escape S.Eur
WHITE'S THRUSH Zoothera dauma – Extremely rare vagrant Sib or C.As
WOOD THRUSH Hylocichla mustelina – Extremely rare vagrant N.Am
HERMIT THRUSH Catharus ustulatus – Extremely rare vagrant N.Am
SWAINSON'S THRUSH Catharus ustulatus – Very rare vagrant N.Am
GREY-CHEEKED THRUSH Catharus minimus – Rare vagrant N.Am
RING OUZEL Turdus torquatus – Migrant
BLACKBIRD Turdus merula – Common breeding resident and migrant
EYE-BROWED THRUSH Turdus obscurus – Very rare vagrant C.As or Sib
BLACK-THROATED THRUSH Turdus ruficollis – Extremely rare vagrant C.As
FIELDFARE Turdus pilaris – Migrant and winter visitor
SONG THRUSH Turdus philomelos – Common breeding resident and migrant
REDWING Turdus iliacus – Migrant and winter visitor
MISTLE THRUSH Turdus viscivorus – Migrant, winter visitor and very rare summer visitor
AMERICAN ROBIN Turdus migratorius – Extremely rare vagrant N.Am

WARBLERS: Sylviidae
CETTI'S WARBLER Cettia cetti – Extremely rare vagrant
GRASSHOPPER WARBLER Locustella naevia – Migrant
AQUATIC WARBLER Acrocephalus paludicola – Rare migrant or vagrant
SEDGE WARBLER Acrocephalus schoenobaenus – Migrant and scarce breeder
PADDYFIELD WARBLER Acrocephalus agricola – Extremely rare vagrant C.As
MARSH WARBLER Acrocephalus palustris – Rare migrant
REED WARBLER Acrocephalus scirpaceus – Migrant and scarce breeder
GREAT REED WARBLER Acrocephalus arundinaceus – Very rare vagrant
OLIVACEOUS WARBLER Hippolais pallida – Very rare vagrant
BOOTED WARBLER Hippolais caligata – Very rare vagrant
ICTERINE WARBLER Hippolais icterina – Scarce annual migrant
MELODIOUS WARBLER Hippolais polyglotta – Scarce annual migrant
DARTFORD WARBLER Sylvia undata – Very rare vagrant
SUBALPINE WARBLER Sylvia cantillans – Rare vagrant
SARDINIAN WARBLER Sylvia melanocephala – Very rare vagrant S.Eur
ORPHEAN WARBLER Sylvia hortensis – Extremely rare vagrant S.Eur
BARRED WARBLER Sylvia nisoria – Rare migrant
LESSER WHITETHROAT Sylvia curruca – Migrant, has bred
WHITETHROAT Sylvia communis – Migrant, breeds occasionally
GARDEN WARBLER Sylvia borin – Migrant
BLACKCAP Sylvia atricapilla – Migrant, has bred
GREEN WARBLER Phylloscopus nitidus – Extremely rare vagrant C.As
TWO-BARRED GREENISH WARBLER Phylloscopus plumbeitarsus – Extremely

rare vagrant C.As
GREENISH WARBLER Phylloscopus trochiloides – Rare vagrant
ARCTIC WARBLER Phylloscopus borealis – Rare vagrant
PALLAS'S WARBLER Phylloscopus proregulus – Rare vagrant C.As
YELLOW-BROWED WARBLER Phylloscopus inornatus – Rare but regular vagrant
RADDE'S WARBLER Phylloscopus schwarzi – Rare vagrant C.As or Sib
DUSKY WARBLER Phylloscopus fuscatus – Rare vagrant C.As or Sib
BONELLI'S WARBLER Phylloscopus bonelli – Rare vagrant S.Eur
WOOD WARBLER Phylloscopus sibilatrix – Scarce migrant
CHIFFCHAFF Phylloscopus collybita – Migrant, summer visitor and scarce breeder, a few winter
WILLOW WARBLER Phylloscopus trochilus – Migrant, summer visitor and scarce breeder

GOLDCRESTS: Regulidae
GOLDCREST Ruglus regulus – Breeding resident and winter visitor
FIRECREST Regulus ignicapillus – Migrant and winter visitor

FLYCATCHERS: Muscicapidae
SPOTTED FLYCATCHER Muscicapa striata – Migrant, summer visitor and scarce breeder
RED-BREASTED FLYCATCHER Ficedula parva – Scarce migrant
COLLARED FLYCATCHER Ficedula albicollis – Extremely rare vagrant
PIED FLYCATCHER Ficedula hypoleuca – Migrant

BEARDED TIT: Timaliidae
BEARDED TIT Panurus biarmicus – Very rare vagrant

LONG-TAILED TIT: Aegithalidae
LONG-TAILED TIT Aegithalos caudatus – Rare vagrant

TITS: Paridae
CRESTED TIT Parus cristatus – Extremely rare vagrant
COAL TIT Parus ater – Occasional winter visitor, birds of European origin formerly bred Tresco and may still do so
BLUE TIT Parus caeruleus – Breeding resident and migrant
GREAT TIT Parus major – Breeding resident and migrant

CREEPERS: Certhiidae
TREE CREEPER Certhia familiaris – Very rare vagrant

PENDULINE TIT: Remizidae
PENDULINE TIT Remiz pendulinus – Very rare vagrant

ORIOLES: Oriolidae
GOLDEN ORIOLE Oriolus oriolus – Scarce migrant

SHRIKES: Laniidae
ISABELLINE SHRIKE Lanius isabellinus – Extremely rare vagrant
RED-BACKED SHRIKE Lanius collurio – Scarce migrant

BIRDS IN THE ISLES OF SCILLY

LESSER GREY SHRIKE Lanius minor – Very rare vagrant
GREAT GREY SHRIKE Lanius excubitor – Very rare vagrant
WOODCHAT SHRIKE Lanius senator – Scarce migrant

CROWS: Corvidae
JAY Garrulus glandarinus – Extremely rare vagrant
MAGPIE Pica pica – Very rare vagrant
NUTCRACKER Nucifraga caryocatactes – Extremely rare vagrant
CHOUGH Pyrrhocorax pyrrhocorax – Former rare vagrant, no recent records
JACKDAW Corvus monedula – Migrant
ROOK Corvus frugilegus – Rare migrant
CARRION CROW Corvus corone – Scarce resident breeder and partial migrant
HOODED CROW Corvus corone cornix (sub-species) – Occasional vagrant
RAVEN Corvus corax – Scarce breeder and occasional visitor

STARLINGS: Sturnidae
STARLING Sturnus vulgaris – Breeding resident, migrant and winter visitor
ROSE-COLOURED STARLING Sturnus roseus – Rare vagrant C.As or M.E

SPARROWS: Passeridae
HOUSE SPARROW Passer domesticus – Breeding resident and partial migrant
SPANISH SPARROW Passer hispaniolensis – Extremely rare vagrant S.Eur or C.As
TREE SPARROW Passer montanus – Scarce winter visitor and migrant

VIREOS: Vireonidae
PHILADELPHIA VIREO Vireo philadelphicus – Extremely rare vagrant N.Am
RED-EYED VIREO Vireo olivaceus – Rare vagrant N.Am

FINCHES: Fringillidae
CHAFFINCH Fringilla coelebs – Breeding resident and migrant, some may winter
BRAMBLING Fringilla montifringilla – Migrant
SERIN Serinus serinus – Rare vagrant
GREENFINCH Carduelis chloris – Breeding resident and migrant
GOLDFINCH Carduelis carduelis – Breeding resident and migrant
SISKIN Carduelis spinus – Migrant and occasional winter visitor
LINNET Carduelis cannabina – Breeding summer visitor and migrant
TWITE Carduelis flavirostris – Extremely rare vagrant
REDPOLL Carduelis flammea – Migrant
ARCTIC REDPOLL Carduelis hornemanni – Extremely rare vagrant
CROSSBILL Loxia curvirostra – Scarce vagrant, any time of year
COMMON ROSEFINCH Carpodacus erythrinus – Rare migrant
BULLFINCH Pyrrhula pyrrhula – Migrant and former breeding resident
HAWFINCH Coccothraustes coccothraustes – Scarce migrant

AMERICAN WOOD WARBLERS: Parulidae
BLACK AND WHITE WARBLER Miniotilta varia – Extremely rare vagrant N.Am
NORTHERN PARULA Parula americana – Very rare vagrant N.Am
MAGNOLIA WARBLER Dendroica magnolia – Extremely rare vagrant N.Am
YELLOW-RUMPED WARBLER Dendroica coronata – Very rare vagrant N.Am
BLACKPOLL WARBLER Dendroica striata – Rare vagrant N.Am
NORTHERN WATERTHRUSH Seiurus noveboracensis – Very rare vagrant N.Am

YELLOWTHROAT Geothlypis trichas – Extremely rare vagrant N.Am
HOODED WARBLER Wilsonia citrina – Extremely rare vagrant N.Am

TANAGERS: Thraupidae
SCARLET TANAGER Piranga olivacea – Extremely rare vagrant N.Am

BUNTINGS: Emberizidae
LAPLAND BUNTING Calcarius lapponicus – Migrant
SNOW BUNTING Plectrophenax nivalis – Migrant
PINE BUNTING Emberiza leucocephalos – Extremely rare vagrant
YELLOWHAMMER Emberiza citrinella – Scarce migrant
CIRL BUNTING Emberiza cirlus – Very rare vagrant
ORTOLAN BUNTING Emberiza hortulana – Scarce migrant
RUSTIC BUNTING Emberiza rustica – Rare vagrant
LITTLE BUNTING Emberiza pusilla – Rare vagrant
YELLOW-BREASTED BUNTING Emberiza aureola – Extremely rare vagrant
REED BUNTING Emberiza schoeniclus – Scarce migrant
[RED-HEADED BUNTING Emberiza bruniceps] – Not accepted to UK list, records considered escape from captivity, extremely rare vagrant
BLACK-HEADED BUNTING Emberiza melanocephala – Very rare vagrant
CORN BUNTING Miliaria calandra – Very rare migrant and former breeding resident
[PAINTED BUNTING Passerina ciris] – Not accepted to UK list, record considered escape from captivity, extremely rare 'vagrant' N.Am
ROSE-BREASTED GROSBEAK Pheucticus ludovicianus – Very rare vagrant N.Am

AMERICAN ORIOLES, BOBOLINK ETC: Icteridae
NORTHERN ORIOLE Icterus galbula – Very rare vagrant N.Am
BOBOLINK Dolichonyx oryzivorous – Very rare vagrant N.Am

LATIN NAMES OF FLOWERING PLANTS
MENTIONED IN THE TEXT

Adder's-tongue fern, small	*Ophioglossum azoricum*
Agapanthus	*Agapanthus praecox*
Arum lily	*Zantedeschia aethiopica*
Autumn lady's tresses	*Spiranthes spiralis*
Babington's leek	*Allium babingtonii*
Balm-leaved figwort	*Scrophularia scorodonia*
Bear's breeches	*Acanthus mollis*
Belladonna lily	*Amaryllis belladonna*
Bell heather	*Erica cinerea*
Bermuda buttercup	*Oxalis pes-caprae*
Biting stonecrop	*Sedum acre*
Blinks	*Montia fontana*
Bluebell	*Endymion non-scriptus*
Bog pimpernel	*Anagallis tenella*
Bramble	*Rubus fruticosus*
Broad-leaved dock	*Rumex obtusifolius*
Butcher's broom	*Ruscus aculeatus*
Button daisy	*Chrysocoma coma-aurea*
Chamomile	*Chamaemelum nobile*
Chicory	*Cichorium intybus*
Chrysocoma	*Chrysocoma coma-aurea*
Common bird's-foot	*Ornithopus perpusillus*
Common bird's-foot trefoil	*Lotus corniculatus*
Common centaury	*Centaurium erythraea*
Common figwort	*Scrophularia nodosa*
Common fleabane	*Pulicaria dysenterica*
Common fumitory	*Fumaria officinalis*
Common honeysuckle	*Lonicera periclymenum*
Common mallow	*Malva sylvestris*
Common poppy	*Papaver rhoeas*
Common ragwort	*Senecio jacobaea*
Common reed grass	*Phragmites australis*
Common rest-harrow	*Ononis repens*
Common scurvy-grass	*Cochlearia officinalis*
Common storksbill	*Erodium cicutarium*
Corn marigold	*Chrysanthemum segetum*
Creeping forget-me-not	*Myosotis secunda*
Cretan mallow	*Lavatera cretica*
Curled dock	*Rumex crispus*
Dwarf mallow	*Malva neglecta*

163

Dwarf pansy	*Viola kitaibeliana*
English catchfly	*Silene gallica*
English stonecrop	*Sedum anglicum*
Escallonia	*Escallonia macrantha*
Euonymus	*Euonymus japonicus*
Eyebright	*Euphrasia officinalis*
Fat-hen	*Chenopodium album*
Fiddle dock	*Rumex pulcher*
Foxglove	*Digitalis purpurea*
Gladden	*Iris foetidissima*
Gladioli	*Gladiolus byzantinus*
Golden rod	*Solidago virgaurea*
Gorse	*Ulex europaeus*
Greater skullcap	*Scutellaria galericulata*
Great mullein	*Verbascum thapsus*
Grey salloy	*Salix cinerea*
Hairy bird's-foot trefoil	*Lotus hispidus*
Haresfoot clover	*Trifolium arvense*
Hedge bindweed	*Calystegia sepium*
Hemlock water-dropwort	*Oenanthe crocota*
Hogweed	*Smyrnium olusatrum*
Hottentot fig	*Carpobrutus edulis*
Ivy-leaved toadflax	*Cymbalaria muralis*
Lady's bedstraw	*Galium verum*
Lesser broomrape	*Orabanche minor*
Lesser celandine	*Ranunculus ficaria*
Lesser hawkbit	*Leontodun taraxoides*
Lesser spearwort	*Ranunculus flammula*
Ling	*Calluna vulgaris*
Long-headed poppy	*Papaver dubium*
Lousewort	*Pedicularis sylvatica*
Male fern	*Dryopteris filix-mas*
Marram grass	*Ammophila arenaria*
Musk storksbill	*Erodium moschatum*
Nettle-leaved goosefoot	*Chenopodium murale*
New Zealand Flax	*Phormium tenax*
Olearia	*Olearia traversii*
Opium poppy	*Papaver somniferum*
Orange bird's-foot	*Ornithopus pinnatus*
Pittosporum	*Pittosporum crassifolium*
Portland spurge	*Euphorbia portlandica*

LATIN NAMES OF FLOWERING PLANTS

Prickly-fruited buttercup	*Ranunculus muricatus*
Prickly saltwort	*Salsola kali*
Primrose	*Primula vulgaris*
Purple loosestrife	*Lythrum salicaria*
Ragged robin	*Lychnis flos-cuculi*
Ragwort	*Senecio jacobaea*
Red clover	*Trifolium pratense*
Red goosefoot	*Chenopodium rubrum*
Red valerian	*Centranthus ruber*
Rock samphire	*Crithmum maritimum*
Rock spurrey	*Spergularia rupicola*
Rosy garlic	*Allium roseum*
Royal fern	*Osmunda regalis*
Scentless mayweed	*Tripleurospermum maritimum*
Sea beet	*Beta maritima*
Sea bindweed	*Calystegia soldanella*
Sea campion	*Silene maritima*
Sea holly	*Eryngium maritimum*
Sea kale	*Crambe maritima*
Sea mayweed	*Tripleurospermum maritimum*
Sea milkwort	*Glaux maritima*
Sea pearlwort	*Sagina maritima*
Sea pink	*Armeria maritima*
Sea purslane (or sea sandwort)	*Honkenya peploides*
Sea radish	*Raphanus maritimus*
Sea rocket	*Cakile maritima*
Sea speenwort	*Asplenium marinum*
Sea spurge	*Euphorbia paralias*
Sea storksbill	*Erodium maritimum*
Sea thrift	*Armeria maritima*
Sheep's sorrel	*Rumex acetosella*
Shore dock	*Rumex rupestis*
Small adder's-tongue fern	*Ophioglossum azoricum*
Small bugloss	*Lycopsis arvensis*
Small-flowered buttercup	*Ranunculus parviflorus*
Smooth hawksbeard	*Crepis capillaris*
Southern marsh orchid	*Dactylorhiza praetermissa*
Spotted medick	*Medicago arabica*
Spring beauty	*Montia perfoliata*
Spring squill	*Scilla verna*
Tamarisk	*Tamarix gallica*
Three-angled leek	*Allium triquetrum*
Toothed medick	*Medicago polymorpha*
Tormentil	*Potentilla erecta*
Trailing St John's wort	*Hypericum humifusum*
Tree mallow	*Lavatera arborea*
Upright St John's wort	*Hypericum pulchrum*

Veronica	*Hebe lewisii*
Viper's bugloss	*Echium vulgare*
Wall oxalis	*Oxalis megalorrhiza*
Wall pennywort	*Umbilicus rupestris*
Water starwort	*Callitriche stagnalis*
Western fumitory	*Fumaria occidentalis*
Wild carrot	*Daucus carota*
Wild mignonette	*Reseda lutea*
Wild thyme	*Thymus drucei*
Wood spurge	*Euphorbia amygdaloides*
Yarrow	*Achillea millefolium*
Yellow bartsia	*Parentucellia viscosa*
Yellow horned poppy	*Glaucium flavum*

RESTRICTED PUBLIC ACCESS TO ISLANDS

Breeding and nesting birds are especially susceptible to disturbance and there are periods when restrictions are imposed on landing on certain of the uninhabited islands within the Park – all of which are held on lease by The Isles of Scilly Environmental Trust.

The following islands are closed during the period 15th March to 20th August, except by written permit:

Annet
The Western Rocks
The Norrard Rocks
Stony Island and Green Island (Off Samson)
Men-a-vaur (Off St Helen's)

Applications for permits for scientific work should be made to
**The Isles of Scilly
Environmental Trust,
St Mary's, Isles of Scilly.
Tel. 0720 422153/422156**

In addition to the above during the period 15th April to 20th July the Islands' boatmen operate a voluntary restriction upon the landing of day trippers on the island of Tean. Other visitors are asked to observe this voluntary restriction in order to avoid disturbing the ringed plovers and terns that nest on beaches above the high water mark.

You are asked to also take heed of certain additional restrictions imposed by The Isles Of Scilly Environmental Trust in respect of all land under its management – this includes all the uninhabited Islands and almost all the coastal areas of the uninhabited Islands, other than the island of Tresco.

Activities which are not permitted on any Trust land are:
Camping
The use of Metal Detectors

Activities which are not permitted on any Trust land other than with prior permission are:

The holding of Barbecues or other lighting of fires.
Shooting.

INDEX

Bold numbers refer to plate numbers. Main reference pages and pages showing sketches are in bold italics.

INDEX

172

INDEX

INDEX